T0362028

Building Limes in Conservation

Building Limes in Conservation

Edited by
IAN BROCKLEBANK

Routledge
Taylor & Francis Group

LONDON AND NEW YORK

First published 2012 by Donhead Publishing Ltd

Published 2015 by Routledge
2 Park Square, Milton Park, Abingdon, Oxon OX14 4RN
711 Third Avenue, New York, NY 10017, USA

Routledge is an imprint of the Taylor & Francis Group, an informa business

© Taylor & Francis 2012

All rights reserved. No part of this book may be reprinted or reproduced or utilised in any form or by any electronic, mechanical, or other means, now known or hereafter invented, including photocopying and recording, or in any information storage or retrieval system, without permission in writing from the publishers.

Product or corporate names may be trademarks or registered trademarks, and are used only for identification and explanation without intent to infringe.

ISBN 13: 978-1-873394-95-3 (hbk)

British Library Cataloguing in Publication Data
Building limes in conservation.
1. Lime. 2. Historic buildings—Conservation and restoration—Materials.
I. Brocklebank, Ian.
691.5-dc23

Library of Congress Cataloging-in-Publication Data
Building limes in conservation / edited by Ian Brocklebank. – First edition.
  pages cm
ISBN 978-1-873394-95-3
1. Lime. 2. Building materials—Conservation. I. Brocklebank, Ian, editor of compilation.
TA434.B89 2012
691'.3--dc23
                2012026129

Typeset by Carnegie Book Production, Lancaster

Front cover photographs: Gerard Lynch, Ian Brocklebank, Ewa Sandström Malinowski, Con Brogan, Grellan Rourke

Contents

Author Biographies

Rosamond (Roz) Artis MA (Hons.), MSc, Dip. Bldg. Cons (RICS), MRICS
Rosamund Artis is the Technical Director of the Scottish Lime Centre Trust (SLCT) and a chartered building surveyor with over 25 years' experience in the building conservation sector, the last 18 years with the SLCT. Her current work includes delivering and developing practical training courses in the use of traditional lime mortars, renders, plasters and limewashes and preparing bespoke building specifications for the repair, conservation and restoration of traditional buildings. Roz is a past committee member of the Building Limes Forum and sits on the Management Committee of the National Heritage Training Group.

Pierre Bergoin
Pierre Bergoin obtained his Diploma of Chemical Engineering at the National College of Clermont Ferrand and graduated in Chemistry at Aix University in Marseilles in 1974.

Since 1981 he has been the Technical Director of St Astier natural hydraulic limes, responsible for production, development and research, where his major task has been the rationalization of a modern manufacturing method capable of achieving constant quality products.

He is a member of the Commission for the revision of the French Standards on Limes and of the Commission for the Standards of Cement. He is also actively involved in various National and International Committees, and has participated in a number of studies and projects related both to ancient mortars and new build.

Geoffrey M. Boffey CChem, MRSC
Geoffrey Boffey was formerly Marketing Development Manager for Steetley Minerals Limited with commercial responsibility for all non-steel lime developments, including dolomitic lime. He is currently proprietor

of CARBOMIN, involved with minerals and lime consultancy, and the trading and distribution of bauxite products.

Ian Brocklebank RIBA, IHBC

Ian Brocklebank is an architect with over 25 years' practical experience in the conservation and repair of all types of historic buildings, including those dating from the Middle Ages to the early Modern Movement. He is a former Editor of the Journal of the Building Limes Forum, subsequently chaired the Forum, and is also a Consultant Editor of the Journal of Architectural Conservation. He now teaches architectural conservation at Plymouth University.

Paul D'Armada BSc (Hons), PGCE

Paul D'Armada joined Hirst Conservation in 1986 having studied and taught earth sciences and physics. As Principal Conservation Scientist, he has significant in-depth understanding of materials science as well as conservation. A specialist in historic plaster and paint analysis, Paul is also a specialist paint technician, advising on the specification of appropriate paint systems (both traditional and modern), together with their production. His research interests include comparative performance of pozzolans, accelerated carbonation of lime and vapour permeability of traditional and modern paint films. His most recent research paper focuses on the innovative technique and use of nano-lime technology.

Alan M. Forster BSc (Hons), PhD, SPAB Scholar, PGCAP

Dr Alan M. Forster is a university lecturer and programme leader for the MSc in Building Conservation (Technology & Management) at Heriot-Watt University, Edinburgh. Alan is a Building Surveyor by profession, an elected board member of the RICS Building Surveying Professional Group (Scotland), and an SPAB Lethaby Scholar (2001). In 2011 he was awarded the Sir Edmund Happold Senior Visiting Research Fellowship within the Department of Architecture & Civil Engineering, University of Bath. He has published many academic journal papers specializing in traditional and low-energy materials, the impact of climate change on traditionally built structures, building resilience and building conservation philosophy.

Steve Foster BSc (Applied Chemistry)

Steve Foster is Business Manager of Singleton Birch Ltd, and Chairman of the Building Limes Forum's Codes and Standards Committee. Steve is a

BSI recognized expert on building limes and chemicals for drinking water treatment. He is a member of the BSI Building Limes Committee, and is the UK representative on the CEN Building Limes Committees. He is a member of the British Lime Association (BLA) Technical and Promotions Committees, and a number of European Lime Association committees. Steve is heavily involved in the development, production and marketing of British Natural Hydraulic Lime, and runs the Singleton Birch's Batts Combe (Cheddar) site.

Alan Gardner BSc (Hons), MRICS, RICS

Alan Gardner specializes in the repair of historic buildings. He was awarded a Society for the Protection of Ancient Buildings Lethaby Scholarship in 1995, worked for them for four years as their Technical Secretary and is a current member of their Technical Panel. After working for English Heritage for a couple of years he returned to private practice, setting up Alan Gardner Associates in 2005.

Elizabeth Hirst ACR, IHBC, FRSA

Elizabeth Hirst is the principal conservator of Hirst Conservation (established 1986). She trained as a stone and medieval wall painting conservator under the auspices of Professor Robert Baker and Mrs Eve Baker. Elizabeth continued her professional development by qualifying as an accredited conservator in 1999. As an architectural conservator, she advises on material consultancy, fine art and historic buildings conservation. She joined the editorial board of the *Journal of Architectural Conservation* in 1995, becoming a consultant editor in 2005. She was also co-editor of *Windows: History, Repair and Conservation.*

Stafford Holmes

Stafford Holmes is an architect specializing in the use of regional building materials and building limes for conservation and sustainable construction.

He became a Partner of Rodney Melville & Partners in 1988, and Senior Consultant from 2006. He is author with others of *Lime and Other Alternative Cements* and *Hydraulic Lime Mortar*; author with Michael Wingate of *Building with Lime*, and author of *Evaluation of Limestone and Building Limes in Scotland* for Historic Scotland and articles for the technical press.

He was Chairman of the Building Limes Forum 2008–2011, is a member of the Society for the Protection of Ancient Buildings Technical Panel and was elected Chairman of the Heritage Skills Hub 2012.

David Hughes BSc (Hons), PhD

David Hughes is Professor of Construction Materials with particular interests in mortars using Roman cements, limes, hydraulic limes and pozzolanas. He has been a prominent member of both the ROCEM and ROCARE EU projects in which his involvement has included calcinations of marls, overseeing a major programme of mortar testing and development of a Standard for ROCARE cements. Particular interests include the influence of sand type and substrate on mortar performance.

Jeremy P. Ingham BSc (Hons), MSc, CEng, CGeol, CSci

Jeremy Ingham is a Principal Materials Engineer with the Special Services Division of Mott MacDonald Ltd. He provides consultancy and investigation services for construction and conservation projects worldwide, involving materials technology, forensic engineering and asset management planning. He has two decades' experience of historic structures and traditional construction materials, and of providing condition surveys, laboratory testing, selection of materials and specification of repairs. He is an expert microscopist and author of the book *Geomaterials Under the Microscope – A Colour Guide*.

Paul Livesey BSc (Tech), EurChem, CChem, MRSC, AMICT.

Paul Livesey is a materials science consultant in cement, lime, mortar and concrete who worked in the cement industry for 40 years becoming Group Chief Chemist and R&D Manager. He liaised with a number of UK and European universities and is currently a visiting research fellow at the University of Bath. He is Chairman of the British Standards Institution Committee for the specification of cements and building limes and a member of the Concrete and Aggregates committees, is a Committee Member of the UK Building Limes Forum and is Chairman of The Scottish Lime Centre Trust.

Gerard Lynch LCG, Cert Ed, MA (Dist), PhD

Dr Gerard Lynch is an internationally acknowledged expert in historic brickwork, master bricklayer, author and lecturer. He is also the author of *Brickwork: History, Technology and Practice* and numerous papers and articles on aspects of historic brickwork and is the leading authority on gauged brickwork. He runs a successful consultancy practice; he also teaches traditional and high-level craft skills, and has worked on and advised on many significant historic brick properties in the UK and overseas.

Grellan D. Rourke B. Arch, RIAI, MICHAWI

Grellan D. Rourke qualified as an architect in 1977 and subsequently undertook postgraduate conservation study in Belgium and Italy. He has worked for the last 34 years in the Office of Public Works in Ireland where he is a senior conservation architect. He does postgraduate lecturing and has published widely on works-related projects. He is a founding director of the Building Limes Forum Ireland, Chair of the Institute for the Conservation of Historic and Artistic Works in Ireland, is President of the Council of ICCROM in Rome and a member of the executive committee of ICOMOS International.

Ewa Sandström Malinowski PhD (Engineering)

Ewa Sandström Malinowski is an architect and senior lecturer and researcher at the Department of Conservation, University of Gothenburg, Sweden. Her main area of interest is the use of traditional materials and crafts in architectural conservation, in particular in the conservation of lime plasters. She coordinated the 'Historic Mortars at Läckö Castle' research project.

Torben Seir Hansen

Torben Seir Hansen is the owner and managing director of SEIR-materialeanalyse A/S, located in Elsinore, Denmark. The company specializes in laboratory analyses and consults on concrete, plaster, lime mortar, masonry, natural stone and surface coatings.

Ugo Spano

Ugo Spano is Director of Setra Marketing Ltd, the technical and commercial liaison office of St Astier for the UK and Ireland. He was a participant in the revision process of the European Standard EN 459:2010.

Vincenzo Starinieri PhD

Vincenzo Starinieri is a Conservation Scientist with particular interests in the conservation of historic buildings and archaeological sites. He has been supporting the ROCARE EU project through the development and evaluation of mortars and processes for a wide range of uses. He has been involved in several conservation projects including the archaeological site of Aigai in Vergina, Greece, and the medieval Abbey of San Martino in Valle in Abruzzo, Italy.

Simon Swann BA (Hons), ACR
Simon Swann is an accredited conservator and has developed a specialist area of interest in historic cements and their use as stucco, cast ornament and decorative surfaces. His work as a consultant has included investigations and advice on Hadlow Tower stucco, Castle House (Bridgwater), Pulhamite Artificial Rockwork at various sites and Felix Austin (and "Austin and Seeley") cast ornament. He is a contributor to the 2012 *Mortars, Renders and Plasters* book published by English Heritage.

Pete Walker BSc, PhD, MIEAust, CPEng, MICE, CEng
Professor Pete Walker, a chartered civil engineer, has been BRE Trust Chair and Director of the BRE Centre for Innovative Construction Materials at the University of Bath since 2006. He has nearly 30 years research experience in the field of structural masonry and materials. His other research interests includes other traditional materials, such as earth building and traditional timber framing, as well as innovative low carbon solutions, such as prefabricated straw bale panels and hemp-lime composites in modern construction. Pete is on the editorial boards of the Institution of Civil Engineer's Construction Materials journal and Building Research & Information journal.

Zhaoxia Zhou BEng, MSc, PhD
Dr Zhaoxia Zhou, a structural engineer, completed her undergraduate and masters studies in civil engineering in China, before completing her PhD on the structural properties of hydraulic lime mortared brickwork at the University of Bath in 2012.

Introduction

Ian Brocklebank

THE DECLINE OF LIME

It has become common to read articles and papers on lime and lime mortars which begin by describing how the material has been in use in building for over 2000 years. These articles and papers often cite Vitruvius, who described ancient Roman use of the material in his *De Architectura* in the first century AD. In fact, the earliest known lime produced specifically for architectural uses dates from the Neolithic period in the Levant, and has been dated to *c.* 10,400–10,000 BC. It is thought the lime was produced in order to make the floors of inhabited caves flat and comfortable for their occupants. An understanding of lime technology, however, seems to be apparent from artefacts dating from around two thousand years before, where the material was used to make sculptures and to plaster skulls for (apparently) ritual purposes. As a human technology, lime therefore predates even the invention of pottery.[1,2] It is interesting to speculate why this might be though it is thought that, before the invention of pottery, nomadic hunter-gatherer people used woven watertight baskets for carrying and cooking goods and liquids. Cooking was carried out by dropping stones heated in the fire into the contents of the basket. If the stone chosen was limestone, it is likely that lime technology would have become obvious fairly rapidly (and rather indigestibly).

Only a few decades ago in the UK and in many developed nations, however, lime was rarely available in the optimum forms for use as the principal binder in building mortars. There were exceptions: a few dedicated individuals caring for buildings of historic importance slaked quick-lime to putty in small quantities, largely for their own use. The majority of informed conservation practitioners used dry, hydrated 'bagged' lime (either with sand for coarse stuff or soaked to a putty) because only this form of the material was commercially available. Portland cement, often mixed with bagged lime, was widely used for building mortars and renders, although on fine works it was sometimes gauged with putty for internal plaster and render on historic buildings.

Within the UK the rise of Portland cement for building mortars had been very rapid since it was first made available commercially in 1845 because its fast set and rapid strength gain both speeded the construction process and provided mortars of unprecedented rigidity and durability. During the later nineteenth and early twentieth centuries, developments in cement chemistry and new production techniques soon enabled rapid increases in both the strength of the material and the scale of overall production. Cement finally became the universal basis for building mortars due to the need for the quick construction of huge numbers of military structures in the Second World War. After 12,000 years as a fundamental human technology, lime had been effectively superseded by Portland cement for all new building works in only a century.

Inevitably, cement came to be the basis for the majority of mortars used for the conservation and repair of historic buildings, and it was sometimes even used neat with what seems now to be a shocking degree of enthusiasm. However, it should not be assumed that all was blind adherence to the wonders of the newer, harder material, or that the knowledge of lime had been lost carelessly. Writing in 1957, Colonel Shore, the technical adviser to the Society for the Protection of Ancient Buildings (SPAB), discusses the historic use of lime mortars, but goes on to recommend the use of cement: lime mortars for actual repairs.[3] His grounds for this were that the industrial atmosphere in Britain's towns and cities of the time contained so much sulfur dioxide that the rain had become dilute sulfuric acid, attacking pure lime mortars, and turning them slowly to water-soluble calcium sulfate (gypsum). Moreover, the resulting sulfation was proving hugely detrimental to softer building stones, particularly limestones, and it was therefore found necessary to reduce the amount of free lime in the associated mortars in order to try to limit the detrimental chemical reactions. The evidence that this was a huge problem at the time is provided by the passing of the Clean Air Act in 1956 (in response to the great smogs of a few years earlier, particularly that of 1952) and by the heavy, black sulfation crusts which we still sometimes find on the uncleaned drying surfaces of urban limestone buildings.

Shore recommended use of what he termed the 'SPAB mortar' as a compromise between 'some of the agreeable qualities of lime mortar and some of the agreeable qualities of cement mortar', and notes that this had also been adopted by the Ancient Monuments Department of the Ministry of Works (now part of English Heritage).[4] The SPAB mortar was a 1:1:6 cement:lime:sand mortar. Shore notes his preference for white cement (for both colour and strength) and coarse sand. The lime would have been hydrated, 'bagged' lime rather than any more active variant of the material, and we would therefore recognize this now as a moderately strong cement mortar in its overall characteristics.

This type of approach became standard procedure for the repair of historic buildings in the UK for the subsequent three or four decades. Shore's mortar specification was still being widely used on historic buildings, particularly for structural repair works, in the early and mid-1990s. The original reason for the adoption of these mortars had ceased to apply and the justification for it had mostly been forgotten. However, there was now simply little alternative to Portland cement for exposed external and wet locations which, with the exception of the small-scale use of pure lime mentioned above, had come to dominate the entire building industry, including the conservation sector.

THE LIME REVIVAL PHASE 1: LIME VERSUS CEMENT

During the 1970s and, more strongly, the 1980s it started to become apparent that widespread use of the standardized cement-based mortars was causing its own kind of problems on historic buildings. Over time, repair works carried out in the preceding decades came to demonstrate a significant range of problems as outlined below.

- It had long been known that mortars should be physically weaker, or softer, than the stones or bricks with which they were used to accommodate differential movement and to be sacrificial in weathering. While it was possible to make relatively soft cement-based mortars, it was not possible to make them sufficiently soft in all cases. This was particularly a problem in parts of the UK where the building materials were themselves particularly soft (as in much of the south-east), but also led to detrimental effects when repairs were being undertaken to already weathered materials.
- Weak cement mortars can have insufficient adhesion to enable them to be used for surface repairs. This can be remedied by making the

mortars stronger but the result will, in many cases, be a mortar that is too hard for the adjoining materials.

- Cement mortars are normally significantly less moisture-permeable than lime mortars, which interferes with the natural processes of moisture diffusion and evaporation within traditional building materials. The introduction of cement can therefore trap moisture within buildings, encouraging decay, and will often force evaporation to take place through the masonry units rather than the mortar joints between them. This transfers erosion patterns to the stone or brick from the more easily replaced mortar.

- The above effect can be considerably exacerbated by the soluble salts that can be present in the building fabric and which will also be introduced by Portland cement. These will dissolve in liquid water but be deposited in the pores of the masonry as the moisture subsequently evaporates. The resulting crystal growth will eventually destroy the cohesion of the material affected, spalling and powdering the surface.

- The rigidity of cement mortars can interfere with the natural movement of traditional structures through thermal expansion and contraction, or by changes in relative humidity. This can result in cracks opening up, which are likely to transport moisture more effectively into structures than normal diffusion. When the cement mortar is particularly strong and well-adhered, the cracks can open within the softer, original material, thus significantly increasing the extent of decay.

- Ordinary Portland cement has a strong dark grey colour, which dominates visually and obscures the aggregate, which gives traditional lime mortars their distinctive appearance. The result can be very harmful to the character of old buildings.

Consequently, while Portland cement-based mortars can have significantly improved durability compared with lime mortars (particularly in severe environments), the qualities that impart this durability are very likely to damage the historic and traditional construction of which they are part. In contrast, the high level of vapour permeability of lime mortars and renders is an essential requirement for the health of traditional solid-walled structures. Buildings based on the use of traditional soft, permeable materials are nearly always more durable overall than those containing a mixture of hard and soft materials, and sometimes more than even the most rigid construction, where small failures and problems can become catastrophic.

The first, and perhaps the most influential, work that underpinned what is now referred to as the 'Lime Revival' in the UK was the conservation of

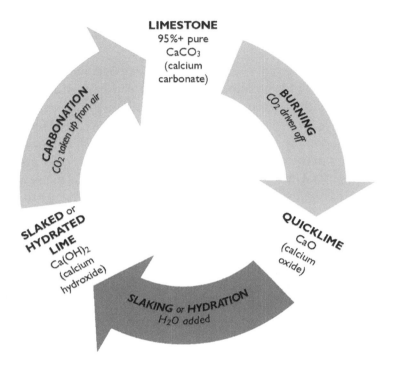

Figure 1   The lime cycle

the West Front of Wells Cathedral in Somerset carried out between 1975 and 1984 by Professor Robert Baker and his wife Eve. The methodology developed here for the conservation of the important medieval limestone statuary, and its associated architectural detailing, has come to be known as 'The Lime Method', and the generation of conservators who were trained by the Bakers include many of those who have become the most highly respected practitioners in their field in the decades since.

The Lime Method relies fundamentally on the lime cycle (Figure 1). When a pure limestone (generally regarded as 95% or more pure calcium carbonate, $CaCO_3$) is burnt at over 900°C, carbon dioxide (CO_2) is driven off, leaving calcium oxide (CaO) – commonly called quicklime because of its strong (sometimes dramatic) reactivity with water. Slaking quicklime is the traditional method of converting this to a binder, where the CaO is mixed with an excess of water to form a putty which is then allowed to mature – preferably for as long as possible to enhance its workability. Modern 'bagged' or builders' lime is hydrated with a controlled minimum amount of water or steam using the correct amount to combine with the calcium oxide without excess, resulting in a dry powder. Both putty and bagged lime, however, are chemically the same substance, calcium hydroxide, $Ca(OH)_2$.

Lime putty is mixed with sand or other aggregate to form a plaster or mortar, and used for building. This process can be replicated by mixing dry hydrate with water and aggregate but, although the chemical results are the same, the physical behaviour of the material tends to be inferior, particularly for conservation work, as the lime will not have had the opportunity to mature and take on the smooth, almost gelatinous quality of a good putty. During the subsequent curing process, which can be slow, the calcium hydroxide reacts with carbon dioxide in the air to return to calcium carbonate. This process is indirect, requiring the presence of water as a medium into which the CO_2 dissolves before it can react with the lime.

The lime cycle is of particular relevance in the conservation of the most valuable sculpture and architectural detail in limestone because it begins and ends with calcium carbonate. This enables conservators to create synthetic limestone for repairs based on lime putty and carefully chosen aggregates. The process is chemically entirely compatible with the objects being conserved and, with skill, repairs can be visually matched with impressive accuracy. This method achieved such success under the Bakers that it has become, with some modifications, the standard method of conserving limestone. Periodic examination of the conserved statues on the West Front at Wells Cathedral indicate that long-term faith in the method, when it has been used with the necessary understanding and skill, is not misplaced.[5]

The diaspora from Wells of those committed to the use of lime spread both knowledge and passion about the material far and wide throughout the building conservation community in subsequent years, both in the UK and internationally. However, it is probably true to say that, due principally to demand and opportunity, a strong concentration of adherents to the Lime Method can still be found plying their trade somewhere near the great Jurassic limestone belt that extends across England from the Dorset coast past Wells, Bath, Oxford and Lincoln as far as North Yorkshire.

As a result, there came to be two principal schools of mortar use: those who persisted with the old cement-based mortars and those for whom only lime was appropriate. On occasions, the lime enthusiasts could become highly averse to cement in any form, creating a dichotomous relationship between the two. Passion for lime was often hugely strengthened by the remarkable elegance of the lime cycle, and the distinctive qualities of lime putty in particular, giving softness, workability and breathability to mortars to a degree that had been almost completely lost in the UK.

While pure putty limes became widely accepted, their use was not entirely straightforward. Although putty provides a wonderful material for the conservation of limestone, and also enabled a resurgence of the craft of lime plastering internally, problems began to arise when putty-based mortars were used in places subject to wet conditions such as for general building work and, more significantly, for external rendering. Failures were found to be occurring in lime putty-based work of all scales, for reasons that were unclear.

It was in response to these twin problems of the continuing widespread use of Portland cement mixes for historic building repairs and failures in pure lime mortars in wet conditions, that the Building Limes Forum (BLF) was founded in 1992 under the guidance of Peter Burman at the Institute of Advanced Architectural Studies at York University. The Forum's stated intention was and remains to 'encourage expertise and understanding in the use of building limes'.[6] Significantly, this was to be a forum of equals where commercial interests and academic rivalries would be put aside and research, discoveries, failures and successes with lime could be freely and openly discussed to the greater good. There have been tensions over the subsequent years, particularly with commercial interests, but this remains the Forum's guiding ethos and a considerable amount of valuable collaborative research has been carried out as a result of this open-minded and forward-thinking stance.

THE LIME REVIVAL PHASE 2: HYDRAULICITY

The overall story remained complex for some time and not all were immediately sympathetic either to the overall virtues of lime, or to those who were trying to understand the material properly or to recover lost skills. Ten years after the founding of the Forum, the early years of the lime revival were caricatured thus:[7]

> Lime plaster has an aura of mystery about it these days, largely because of the Building Limes Forum, a collection of mostly middle-class conservation enthusiasts, who have discovered the joys of working with lime, but insist that the only way is to use traditional lime putty, preferably slaked in oak vats from quicklime fired in medieval kilns, using water gathered by virgins from dewdrops at the midsummer solstice.

There may at one time have been a certain grain of truth in this, but by

2002, both the important issues and the wider level of understanding had advanced considerably. There was by this time a growing body of valuable work that was uncovering a much more complex range of historic mortars and their binders, as well as achieving some significant insights into their use.

Towards the end of the Bakers' project at Wells, in 1983, the Ecclesiastical Architects' and Surveyors' Association published the first edition of John Ashurst's book, *Mortars Renders and Plasters in Conservation*.[8] This relatively thin volume contained a surprisingly wide range of information about the different materials that might be found in old buildings, and put forward valuable techniques for their use and conservation. Overall, perhaps the greatest lesson the book contained was simply a reminder of just how great a variety of materials had historically been available.

Nevertheless, mortars recommended in this important work still included 1:1:6 and 1:2:9 cement:lime:sand mixes in certain circumstances. Subsequent publications, including John and Nicola Ashurst's volume on mortars, plasters and renders in English Heritage's 1988 series, *Practical Building Conservation*, continued to recommend 1:1:6 and 1:2:9 cement:lime:sand mixes for the repair of early cement and proprietary mastic finishes, though there is some discussion of hydraulic limes, reflecting their limited appearance on the market at this time.[9] Interestingly, these are the same mixes recommended by standard construction textbooks and British Standards for new masonry building and rendering during the same period up to the present day.

Probably one of the first, if not the first, publication in the late 1980s that focused on building limes without Portland cement additions was the SPAB Information Sheet, *An Introduction to Building Limes*, by Michael Wingate,[10] which remains an excellent introductory text covering as it does both pure and hydraulic limes. This was followed rapidly by Bruce and Liz Induni's *Using Lime*,[11] *Lime and Other Alternative Cements* by the Intermediate Technology Development Group (ITDG),[12] *Lime in Building* by Jane Schofield,[13] and Holmes and Wingate's *Building with Lime*.[14]

These publications combined over a short space of time to fill a significant gap in general knowledge of historic mortars, and to enable both much wider debate and more confident practical conservation work, to the huge benefit of the entire field. Not only did these works collectively explain best practice in how pure limes should be used, thus reducing the failure rate, but those by Holmes and Wingate and the ITDG publications also set out the place and value of hydraulic limes in historic use, and clarified that it was this class of materials that were optimum for use in weathering environments, rather than the pure limes which were common in the early years of the lime revival. It is immensely disappointing that the production of feebly hydraulic, grey chalk limes at Totternhoe in Bedford-shire, believed to have commenced around 1650, ceased in 1993 just as the value of the material for conservation purposes was being rediscovered. Papers by Stafford Holmes and Ian Brocklebank in this volume make clear the significance of that loss.

At around the same time as the founding of the BLF, and inspired by the same need for better knowledge, was the first stage of English Herit-age's 'Smeaton Project' which set out to understand the variable factors influencing the behaviour and performance of lime mortars.[15] This project gave useful information on the effectiveness of certain pozzo-lanic additives to encourage a set in pure limes, and also the dangers of adding small amounts of cement to lime-based mortars. With hindsight, it is notable how small a part hydraulic limes played at the beginning of this project.

The Smeaton Project was an extremely valuable initiative which contributed to a significantly enhanced understanding of the use of lime mortars and renders in the UK, particularly in weathering environ-ments. However, its success was rapidly followed by the introduction of a range of hydraulic limes into the market. Initially these were from established suppliers in France, but production of a new, high quality, moderately hydraulic lime in England commenced in 1997 using the blue lias rocks of Somerset, which had been made so famous by John Smea-ton's discoveries about hydraulic limes in the mid-eighteenth century. It is of course in homage to these early discoveries that English Heritage named their research project. Unfortunately, the apparent gains in this regard rapidly evaporated. The excellent blue lias lime was forced out of production in 2006 for legal reasons, although there remains a UK manufacturer of hydraulic limes in north Lincolnshire, fortunately with a sufficiently large and varied overall production capacity to be likely to survive for a long time yet.

English Heritage continued the Smeaton Project for two further phases, during which the range of study was widened and adjusted to correspond with changes in the available palette of materials, especially the introduc-tion of hydraulic limes. By 1997, when English Heritage reported on the completion of the third phase of the project, these had become the focus of considerable interest within the conservation world, as can be seen from the contents of the issue of *Lime News* (the precursor of the *Journal of the Building Limes Forum*) published in the autumn of that year.[16] The vast majority of publications on lime from this period and after, such as Holmes and Wingate[17] and the second edition of John Ashurst's book[18] have included the recommendation that, as a wide range of hydraulic limes is now available, there is simply no need to include Portland cement in conservation mortars or renders.

Almost ten years after phase one of the Smeaton Project came the Foresight Lime Research programme, which was based at The Interface Analysis Centre of Bristol University under the direction of Professor Geoff Allen. What the Smeaton Project achieved for air/pure lime, the Foresight research achieved for hydraulic limes. This project researched the performance, application and classification of hydraulic limes for use in building. It examined in detail the functions and properties of hydraulic lime mortars, the constituent materials, workmanship, appearance, strength requirements and all aspects of selecting for durability in various types of exposure conditions. In a nutshell it answered the questions likely to be posed by anyone contemplating the use of hydraulic lime for building mortars, and the finished report remains a standard reference.[19]

While commercially produced hydraulic limes are reliable, consistent, cost-effective and well-supported by their manufacturers, and therefore well-liked and widely used in conservation and the wider building industry, there are important aspects of their manufacture and specification that need to be understood. Many of these arise from the history and form of the European Standard for building limes which governs production and specification. In 2001, the British Standard for building limes BS 890 was superseded by the European (CEN) Standard BS EN 459:2001, which has now been revised and reissued as BS EN 459:2010 (this is identical to EN 459:2010). The changes brought about by these modifications in standards and their implications for practitioners are discussed in detail in this volume.

Figure 2 demonstrates diagrammatically, in a simplified way, the four principal families of limes and cements which are now generally recognized and referred to based on the different ways in which they are used and produced. These processes are themselves a product of the various chemistries at work. The inner circle is the pure lime cycle, as shown in Figure 1. Outside this is the natural hydraulic lime family, which contains both pure lime and a hydraulic set based on impurities in the parent limestone. Natural (often called 'Roman') cements are made from even more impure limestones that contain large amounts of aluminates, giving these materials a very fast set, while the modern, artificial Portland cements are made from mixtures of limestones and clays, and fired at

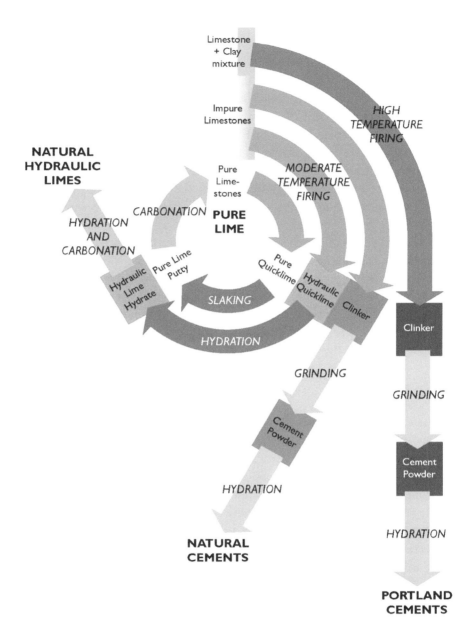

Figure 2   Four principal families of limes and cements.

higher temperatures for yet another chemical result. Further information concerning all these families, and their importance for various types and ages of buildings, can be found in this book.

CONTINUING RESEARCH

At the time of writing, there is a sufficiently wide range of materials available on the UK market that the use of Portland cement has been substantially eliminated from the field of architectural conservation, at least on pre-twentieth-century buildings. Pure lime putty and the range of natural hydraulic limes defined in BS EN 459:2010 (NHL2, NHL3.5 and NHL5) are now generally considered to cover most eventualities – and this is the situation in many cases.

However, research has continued within the BLF and elsewhere, and there remain causes for concern. It appears that often we have been premature in assuming that we have all, or enough, of the answers and we have discovered that many aspects of the use of lime require further understanding. There are, in addition, still important gaps in the range of materials needed to properly conserve all valuable buildings.

This book contains a range of papers, mostly from the *Journal of the Building Limes Forum* and the *Journal of Architectural Conservation*, which have been instrumental in progressing knowledge of these important issues in the field over recent years.

New understanding of historic materials

The material in Part 1 gives new and valuable insights into the nature and origins of historic building limes and lime-based mortars. The first paper, 'The Lime Spectrum' by Ian Brocklebank (page 21), is a distillation and combination of understanding, obtained from scientific testing, historic literature and practical use, which has been built up by members of the BLF. It is included here as an overview of the full range of materials and their interrelationships as we now understand them.

Pierre Bergoin's paper on 'Lime after Vicat' tells the story of the developing understanding of hydraulic limes, particularly in France, from Louis Vicat's work in the early nineteenth century to the publication of BS EN 459:2010, the current European standard classifying building limes in general (page 33). It provides a crucial understanding of the reasoning behind the current standard and describes the background to the French hydraulic limes so prevalent in the international market.

Pete Walker and Zhaoxia Zhou discuss the factors affecting the structural strength of masonry built with lime mortars (page 53). Their work is based on comprehensive laboratory testing, using methods which allow direct comparison with cement-based mortar standards. Of particular importance is the different pattern of strength gain in limes as opposed to cements, which it is vital to understand when specifying mortars to be used for structural purposes. It is particularly instructive to read this paper in conjunction with the one by Paul Livesey on 'Portland Cement Properties Through the Ages' (page 65), which usefully explains the reasons why the strength of artificial cements have so dramatically increased over the past 150 years. Clearly, Portland cements of the past are significantly different to those we purchase in builders' merchants today and are even less suitable for use in maintaining historic buildings than those advocated by Colonel Shore 50 years ago.

Ian Brocklebank sets out the story of the developing use of lime in London through a critical phase in the urban development of the city, relating documentary evidence to architectural history through practical observation (page 73). The financial and practical pressures of building in the capital mean that lime is now probably less understood and appreciated in London than almost anywhere else in the UK. Where pure limes and slightly hydraulic lime were once commonplace, the tendency now is to specify over-strong mortars for speed of set.

Finally in this section, David Hughes, Simon Swann, Alan Gardner and Vincenzo Starinieri discuss recent studies into Roman cements both within the UK and in Europe (page 107). This family of materials dominated the market in the first half of the nineteenth century for durable mortars suitable for rendering and modelling the exteriors of many buildings of the time, giving rise to the Regency style of architecture as well as contributing to the early Gothic revival and some fascinating developments in new forms of construction. The link between this material and the changing architectural language of the time is critical and it has always been very difficult to repair Roman cement with other materials from the lime spectrum. The renewed production and increased understanding of this material should be strongly welcomed.

Materials analysis, testing and standards

2010 marked the introduction of the first revision of BS EN 459 *Building Lime*. The Building Limes Forum was pleased to be consulted on its development although only a few, but significant, recommendations by the Forum were finally incorporated. The revision contains some valuable changes and these are discussed in an introduction by Steve Foster (page 133). However, the 2010 version remains imperfect in some important ways and Stafford Holmes, in his piece 'To Wake a Gentle

Giant – Grey Chalk Limes Test the Standards' (page 199) sets out the background to a significant problem. This is the lack of any category of material matching the feebly hydraulic, grey chalk limes that were historically so prevalent in the south-east of England (including London) and highly respected further afield. It is a major omission that the categories of natural hydraulic lime (NHL) mandated in the 2010 standard allow at their weakest only for a material (NHL2) which is equivalent to a *moderately* hydraulic historic lime. Although there may be only specific uses for a weaker hydraulic lime in new buildings, the fact that a *feebly* hydraulic lime is effectively prohibited by the standard is unnecessarily restrictive to conservation. Perhaps the use of the word 'feebly' in this instance seems pejorative (it is a literal translation of a term coined in French by Vicat), but the material offers a mix of desirable qualities unmatched by any other. In particular, it is the most suitable form of lime for gauged brickwork and fine ashlar masonry.

Roz Artis of the Scottish Lime Centre discusses sands for use with lime mortars (page 145). This is an area that could perhaps have benefitted from greater discussion. So much attention has been paid over the past 20 years or more to understanding lime types that it is easy to forget that the greater part of most mortars is aggregate and that this can sometimes be even more important than the binder for deciding vital qualities such as compressive strength and vapour permeability.

Jeremy Ingham provides two papers on the testing and assessment of lime mortars, one covering laboratory testing and the other testing for failures in the field (pages 155 and 175). This is an area that has been badly neglected by most publications in the past, but is immensely important to conservation. In particular, the range of possible laboratory tests described here will give those faced with difficult or obscure historic mortar samples the knowledge to usefully discuss and commission the testing methods possible to obtain the information they need. There will, however, always be the requirement for careful and knowledgeable interpretation, as laboratory test results are notoriously difficult to translate into practical mortar specifications.

REDISCOVERING METHODS AND TECHNIQUES

The papers in this section include valuable new understandings about the ways in which mortars were made and used historically, which has the potential to significantly change current practices. If we choose to listen and observe, we have the potential to better match historic building mortars and to achieve much greater degrees of success in their use than is possible using the materials that are commercially dominant today.

Gerard Lynch's paper 'Lime Mortars: The Myth in the Mix' points out an important basic truth: that historic mortars were often significantly richer in lime than we usually recognize (page 221). He explains with considerable clarity, and from a position of long practical experience, that 1:3 mortars made from putty are far weaker than those made directly from quicklime, as was normal practice. There was no economic advantage in using an expensive and rare putty lime to mix with sharp sand and grit for general building mortars when quicklime mixed directly with damp sand actually produced a better result. When this method is combined with Stafford Holmes' recognition of the historical importance of grey chalk limes, some important truths about building with lime start to emerge and it is to be hoped that these will be recognized more widely.

The technique of mixing quicklime directly with damp sand is usually called 'hot mixing' or 'hot lime', and its effects on the qualities of the final resulting mortars are explored in considerable detail by Alan Forster. This became a seminal piece, being directly inspired by the work undertaken by Ewa Sandström Malinowski on the substantial re-rendering of Läckö Castle in Sweden, which is included here in a detailed case study (page 271).

Despite the widespread availability of proprietary natural hydraulic limes, there is still a place for pozzolanic additives in conservation. Not only is it important to understand their effects in order to understand and replicate historic mortars, but they may still offer useful qualities, especially where a particular strength or set is required, or where it is desirable to limit the leaching of free lime in an otherwise weak mortar. Geoff Boffey, Elizabeth Hirst and Paul Livesey provide a valuable paper on these materials, setting out their variety and behaviour, as well as including some valuable notes of caution to assist their successful use if desired (page 229).

This section concludes with two papers of direct relevance to stone and plaster conservation. Grellan Rourke's paper on the analysis and replication of the historic mortars at Ardfert Cathedral in Ireland is a case study on the high-quality conservation of medieval masonry, carried out and recorded to exemplary standards (page 295).

Finally, Elizabeth Hirst and Paul D'Armada have provided a paper that introduces nano-lime and discusses the results of recent tests into its use for the conservation of stone and plaster (page 321). One of the more contentious aspects of the Bakers' original development of the lime method at Wells Cathedral was their recommendation of the application of lime water for the consolidation of friable, powdery stone surfaces. Lime is only very mildly soluble in water so, in order to have any useful effect, the limewater needed to be applied many times. Sometimes hundreds of applications were required and the consolidation effects might still be negligible, while the water could easily mobilize salts or cause other damaging side effects. Nano-lime, in contrast, is a colloidal suspension of calcium hydroxide in alcohol, which has the potential to deliver far greater quantities of lime far deeper into any friable material, for greater consolidation effect, while avoiding the introduction of damaging water. This is an entirely new material and it is still very much at the testing stage. However, interest has spread widely since the first rumours of its potential and it is hoped that it may offer very real benefits for the future.

THE CONTINUING STORY

This book encapsulates a large proportion of the most important research work carried out in the field of lime mortars in conservation between 1995 and 2012. It should not, however, be assumed that the overall research project has been completed. The Building Limes Forum and others are actively continuing their quest for further understanding.

There remain deficiencies in the current iteration of BS EN 459, which have widespread and detrimental consequences on the specification and production of the optimum mortars for historic building conservation. BLF members are continuing their constant struggle to revise both British and European standards to pay due attention to British building limes and give confidence to specifiers and users.

A further project is being developed at Bath University, also initiated by the BLF, to examine the varying properties and performance of the wide variety of hydraulic limes currently on the market. This latest piece of proposed research, together with the Smeaton and Foresight Projects, are vital for giving credibility and confidence in the use of lime, both for repairing the historic estate and for new building works. The precise physical characteristics and levels of performance of building limes for the satisfactory conservation and repair of historic buildings are not covered sufficiently in BS EN 459:2010. The Standard was written primarily by and for lime producers, and the classifications given are too broad for specifiers, particularly where the conservation of delicate building fabric is the priority. It is therefore important to establish the degree of variation between the building limes produced by different manufacturers.

Although the Building Limes Forum was originally founded to research and understand materials within the conservation field, those materials are now experiencing a huge resurgence within new building, where lime has been found to be of particular significance in response to the need for sustainability. It is particularly gratifying that limes are now sought for the construction of high-quality new buildings on a significant scale due to their particular physical and chemical properties, and not simply for romantic reasons. At least half of the overall interest and research into limes within the BLF will shortly be devoted to new build projects, a situation which could only have been a wild dream for its founders.

As an organization, the Forum has also been a significant success due primarily to its original guiding ethos of cooperation and a series of sister organizations that have grown up in different parts of the globe. The BLF maintains very close and cordial links with the independent Building Limes Forum Ireland (www.buildinglimesforumireland.com) and still provides a focus for much of the rest of the English-speaking world. It also works closely with the Nordisk Forum for Bygningskalk (www.kalkforum. org) in Scandinavia, which has carried out much good work over many years, and the Forum Italiano Calce (www.forumcalce.it), which seeks with immense enthusiasm and drive to improve the conservation of the huge number of valuable historic buildings in Italy. At the time of writing, new forums are under discussion in Spain and India, and there is active interest in Australia, France and South America. The Building Limes Forum will continue to do all it can to assist these efforts and anywhere else in the world where the need may be felt.

Acknowledgements

Many thanks are due to Stafford Holmes for valuable insights, comments and clarifications in the above text.

Notes

1 Gourdin, W. H. and Kingery, W. D., 'The beginnings of pyrotechnology: Neolithic and Egyptian lime plaster', *Journal of Field Archaeology*, Vol. 2, No. 1–2, 1975, pp. 133–50.

2 Kingery, W. D., Vandiver, P. D. and Prickett, M., 'The beginnings of pyro-technology. Part II: Production and use of lime and gypsum plaster in the pre-pottery Neolithic Near East', *Journal of Field Archaeology*, Vol. 15, No. 2, 1988, pp. 219–44.

3 Shore, B. C. G., *Stones of Britain: A Pictorial Guide to those in Charge of Valuable Buildings*, Leonard Hill (Books), London, 1957.

4 Ibid.

5 Durnan, N., 'Wells Cathedral: West Front 2002. Report on the re-treatment of statues', *Journal of the Building Limes Forum*, Vol. 10, 2003, pp. 54–63.

6 Building Limes Forum website, www.buildinglimesforum.org.uk

7 Howell, J., 'On the level; plastered', *Daily Telegraph*, 12 June 2002.

8 Ashurst, J., *Mortars, Renders and Plasters in Conservation*, Ecclesiastical Architects' and Surveyors' Association, London, 1983.

9 Ashurst, J. and Ashurst, N., *Practical Building Conservation, Volume 3: Mortars, Plasters and Renders*, English Heritage Technical Handbook Series, Gower Technical Press, Aldershot, 1988.

10 Wingate, M., *An Introduction to Building Limes,* Information Sheet 9, The Society for the Protection of Ancient Buildings, London, 1989.

11 Induni, B. and Induni, E., *Using Lime*, Taunton, Somerset, 1990.

12 Hill, N., Holmes, S. and Mather, D., *Lime and Other Alternative Cements*, Intermediate Technology Publications, London, 1992.

13 Schofield, J., *Lime in Building*, Black Dog Press, Cullompton, Devon, 1994.

14 Holmes, S. and Wingate, M., *Building with Lime*, Intermediate Technology Publications, London, 1997.

15 Teutonico, J. M., McCaig, I., Burns, C. and Ashurst, J., 'The Smeaton Project: factors affecting the properties of lime-based mortars', *APT Bulletin*, Vol. 25, No. 3–4, Association for Preservation Technology International. A shortened version can be found in *Lime News*, Vol. 2, No. 2, June 1994, Building Limes Forum.

16 *Lime News*, The Building Limes Forum, Vol. 5, 1997, pp. 38–86.

17 Holmes and Wingate, *op. cit.*

18 Ashurst, J., *Mortars, Renders and Plasters in Conservation*, 2nd edition, Ecclesiastical Architects' and Surveyors' Association, London, 2002.

19 Allen, G., Allen, J., Elton, N., Farey, M., Holmes, S., Livesey, P. and Radonjic, M., *Hydraulic Lime Mortar for Stone, Brick and Block Masonry*, Donhead Publishing, Shaftesbury, 2003.

New Understandings
of Historic Materials

The Lime Spectrum

Ian Brocklebank

The building conservation community has for many years been fully familiar with the importance of lime in the repair and maintenance of historic buildings of all ages, and the danger of using cements inappropriately. Unfortunately, this is still often regarded as a simple dichotomy between two opposed materials: lime = good; cement = bad. In reality, however, pure lime and modern ordinary Portland cement (OPC) lie at the two extremes of a broad spectrum of materials with essentially allied characteristics, and the materials which are most appropriate for use on any particular historic building may well be found from anywhere within this range.

The diagram shown in Figure 1 is a simplified representation of the spectrum of limes and cements which have historically been available. It is set against a scale of the range of compressive strengths that can generally be achieved (in MPa, which are the same as N/mm^2) with the weakest to the left and the strongest to the right. This scale demonstrates very clearly the huge difference between historic limes as found on most old buildings, and the extraordinary strength and rigidity which modern OPCs can achieve. This alone illustrates that these are materials which, despite being within the same generic family, have vastly different physical properties and are clearly therefore intended for very different purposes.

Compressive strength is by no means the only method by which limes and cements can be classified, and in conservation work it is probably not even the most important quality, but it has the virtue of allowing convenient comparison between a wide range of materials and also relates directly to current standards.

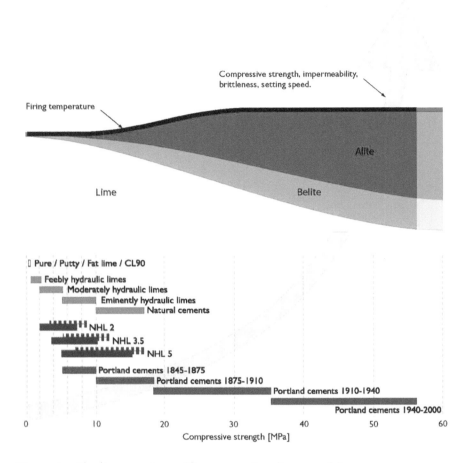

Figure 1   The lime spectrum (diagrammatic – linear scale).

Note that the compressive strengths quoted here are derived from laboratory testing using processes standardized for material comparison only and do not reflect the compressive strengths likely to be desirable in actual conservation mortars, where a far wider range of variables will always apply. The data for the older classifications of both limes and cements have been derived from historical documents and standards, generally using imperial units, and converted to suit. This is unlikely to provide a completely accurate comparison with the modern materials as historic testing methods differed, but it is close enough to be within an acceptable margin for error for most practical purposes.

CHEMISTRY AND MATERIALS

Air limes

The different lime types are defined by the principal setting chemistries at work, as it is from this factor that their various qualities arise. Pure lime (shown in light grey in Figure 1) sets by the carbonation of calcium hydroxide ($Ca(OH)_2$) to calcium carbonate ($CaCO_3$, the principal material of limestone) through a slow reaction with atmospheric carbon dioxide (CO_2). This reaction, however, can only take place in the presence of water, which is why slow curing is vital. For the most effective result the mortar, plaster or render needs to be kept damp for weeks to enable the reaction to take place thoroughly. A balance is essential, as too rapid drying will lead to incomplete carbonation, but excessive wetting will cause saturation of the pores in the material and prevent the carbon dioxide reaching beyond the surface. This necessity for dampness is the reason lime work should not normally be carried out when there is any danger of frost.

Materials which cure by carbonation alone are pure limes or air limes (at the left end of the spectrum), and are generally available either as lime putty or as bag lime from builders' merchants (CL90 to European standard BS EN 459-1:2010). These materials are chemically almost identical, both being calcium hydroxide, but their different methods of production give them different physical properties.

In a putty, the calcium hydroxide is produced by slaking quicklime (calcium oxide, CaO) with an excess of water and then allowing it to mature over time, while bag lime is produced industrially by hydration using exactly the correct amount of water for the amount of CaO. The maturing process in a putty allows the calcium hydroxide to break down slowly and thoroughly to achieve the characteristic smoothness, workability and stickiness of a fine putty, while hydrated lime is a dry product sold in paper bags, which can give a powdery or slightly granular texture to mixes. The latter may not be a problem when mixed with sharp sand and cement on a modern building site, but it should not really be used for fine masonry jointing, plasterwork or limewash, for which putty is ideal.

Hydraulic limes

Hydraulic limes and cements are produced from mixtures of limestone with clays. This may occur naturally in impure limestones (giving rise to natural hydraulic limes, NHLs in BS EN 459-1). Alternatively it can be achieved by deliberate addition in measured quantities before firing (hydraulic limes (HLs) in the same standard, although these are also permitted to include a wide range of other chemistries, not all of which will be desirable for conservation use). The clays, when activated by the firing, combine with the water and the free lime in the mix to produce a set primarily based on hydrated calcium silicates. Critically, this reaction can occur in the absence of air, such as under water, giving rise to the terms 'hydraulic lime' and 'water lime'. In general, the greater the proportion of clay in the original material, the stronger is the hydraulic set that results.

There are two principal types of hydraulic component, imaginatively christened 'alite' and 'belite' by cement chemists. Alite (shown in dark grey in Figure 1) is composed primarily of tricalcium silicate (C_3S), while belite (shown in mid-grey in Figure 1) is based on dicalcium silicate (C_2S). Both types of hydraulicity can be derived from the same initial material, but the active compounds that result depend on the firing temperature, with alite only being produced above 1260°C. Belite is formed at a range of temperatures between (roughly) 900 and 1200°C, which is the normal range of firing temperatures for limes. Alite sets harder and faster than belite, and gives cement its particular qualities, while hydraulic limes are characterized by a set based primarily on belite.

Hydraulic limes always consist of a mixture of lime and belite acting in parallel. This dual curing process should be taken into account whenever they are used, ensuring that both the initial hydraulic set and the slower carbonation of the free lime can take place, though for practical purposes the latter may now be less critical. When the proportion of belite is low, the mix is more workable and cures slowly and softly to give a very permeable and flexible finished material. A higher proportion of belite gives a faster and harder set at the expense of workability, permeability and flexibility. Carbonation continues to follow the hydraulic set at its natural pace until the material reaches its finished, cured state, and so protection and tending are required on-site with even a moderately strong hydraulic lime.

Aluminates are also normally present in smaller quantities as part of the secondary chemistry. These contribute to the overall hydraulicity and are the most reactive components in the material, usually imparting a particularly quick set if present in significant quantities (as is the case in most historic 'Roman' cements). Ferrite phases can also form a relatively large proportion of the overall chemistry. Both are omitted from Figure 1 to aid clarity. Detailed consideration of these particular components is only necessary in unusually critical situations, in which case the advice of specialists should ideally be sought.

Pozzolans

Pozzolans are separate activated silicates, usually derived from fired clays, either from brick or other industrial production, or naturally fired in volcanoes. These increase the hydraulicity of a mix by combining with the free lime during the curing process to form belite (ground brick dust and natural volcanic earths, such as those from Pozzuoli near Naples and trass from Germany) and sometimes even alite (ground granulated blastfurnace slag, pulverized fuel ash and similar high temperature materials). Thus any material containing a proportion of free lime can be moved to the right on the spectrum shown in Figure 1 by the addition of suitable pozzolanic additives.

Modern cements

Modern ordinary Portland cement (OPC) contains belite and lime (calcium hydroxide is actually produced as a by-product during cement hydration), but the speed and strength of the alite set is deliberately designed to swamp the effects of the two weaker, slower processes. Such a material is extremely well-suited to the production of modern reinforced concrete, giving quite remarkable material densities and compressive strengths, and it can be used with little protection or curing on site, making it very tolerant of poor workmanship.

In order to improve the economy and environmental sustainability of their production, modern OPCs now tend to be manufactured from a wide range of extracted and waste materials. Complex chemistry is thereby introduced which is often undesirable for conservation purposes and about which the manufacturers are not always entirely open. Gypsum is also added to retard the speed of the natural set (which can otherwise take only seconds). This results in a material that is not only unnecessarily hard, brittle and impermeable but which also introduces a significant load of soluble sulfates. This can cause efflorescence on facework when these are mobilized by water, and severely degrade adjacent brick and stone by re-crystallizing within their pores. The latter is a very common failure mechanism where modern concrete or cement has been introduced into historic masonry. Most mass-produced bricks are now manufactured with an exaggeratedly high resistance to salt damage, partly in order to cope with the cement, and this is a main reason for the difficulty in matching them with their historic softer-fired predecessors.

A range of proprietary additives (including retarders, plasticizers, water-retaining and air-entraining additives) have been developed in the past 50 years or so to modify the performance of OPC, at least partly to try to make it more suitable for masonry use. However, all these also add potentially undesirable chemistry for conservation work.

Although OPC is an impressive material, which is convenient and economic to use, it is not really appropriate for modern masonry construction (finished work requires copious use of expansion and movement joints), is inevitably of inferior quality, and is usually quite ugly. It is also almost completely unsuited for use in the conservation field. Moreover, the various grades of hydraulic limes can normally produce exactly the qualities required without introducing either undesirable physical behaviour or any unpredictable chemistry.

HISTORIC AND MODERN LIMES

Historic and modern classifications

A tripartite classification of hydraulic limes into 'feebly', 'moderately' and 'eminently' hydraulic bands, based on a modification of Vicat's original system of 1818, was widely adopted within the UK until the demise of most of the traditional lime industry as it was both practical and appropriate to the end uses of the materials. This classification has now been superseded by the natural hydraulic lime classifications of NHL2, 3.5 and 5 enshrined in the European standard BS EN 459-1:2010.

However, these two tripartite classifications are not equivalent. The weakest modern class of NHL2 is more closely equivalent to the traditional 'moderately hydraulic' class, while NHL3.5 is nearer to 'eminently hydraulic'. NHL5 limes can easily achieve strengths equivalent to natural cements. As can be seen in Figure 1, there is now a critical divergence in the modern classification between pure limes and NHL2 where the traditional 'feebly hydraulic' class used to comfortably sit.[1]

This disparity is particularly important within the UK as many of the limes which were highly regarded historically, and greatly sought after, were classed as 'feebly hydraulic'. These stone limes (as opposed to chalk or pure putty lime) are also a particularly interesting category for their working characteristics. Whereas modern NHLs are invariably produced like bag lime by hydration, and sometimes even need to be ground like cements, the feebly hydraulic limes were traditionally delivered to site as quicklime and mixed directly with the sand, before being left to finish

slaking in the mix. This gave a mortar with the valuable combined qualities of the workability of a putty-based mix (without the expense, delay
and effort) and the convenience of a slow hydraulic set. Indeed it is this
mix of qualities for which the feebly hydraulic limes were so highly prized.
It is disappointing that BS EN 459-1:2010 omits this class, not because
there is no use for these limes, but because the test method used to assess
compressive strength (the crushing of a standard cube) is derived from
cement testing methods and is simply unreliable below a crushing strength
of 2 MPa. These limes could also be run to putties for fine-gauge brickwork
or plaster details, in which case they would retain both their workability
and hydraulic properties for perhaps several weeks.

The spectrum also contains a little recognized difference between
historic and modern hydraulic limes, which can on occasion be significant
in conservation work. All modern NHL limes are produced by industrial
hydration to form powders, usually sold in paper bags. This is a particularly significant technology, as the process enables just the right amount
of water to be added (in the form of steam) to the burnt lime to slake the
CaO to $Ca(OH)_2$, while not adding enough to trigger the belite to start
its hydraulic set. This works because the lime is far more reactive than
the belite and so takes up the available water much faster. However, prior
to this technology becoming available towards the end of the nineteenth
century, it was not possible to calibrate the correct amount of water with
sufficient accuracy. Hydraulic limes would therefore either be slaked with
too much water, causing a set to start before it was required, or with too
little, in which case there would be unslaked lime in the material which
could react after placing, thus ruining the work. In practice, this limited
useful hydraulic lime production to those types lying at the weaker end
of the spectrum, where the hydraulic set was slow enough to be manage-
able. Where stronger limes were required for civil engineering and similar
works, these would normally be produced by adding a significant quantity
of a pozzolan perhaps to a moderately hydraulic lime. The stronger NHLs
shown here, therefore, are almost exclusively modern materials, although
many of them are more convenient for conservation purposes than the
historic equivalent mixes would have been.

Lean limes

Lean limes are frequently mentioned in historic sources. The terms 'rich'
and 'lean' when applied to limes are actually craftsmen's terms and refer
to the amount of sand which a particular lime can usefully carry, rich

limes being able to hold more than lean ones. Although not generally categorized as such, hydraulic limes also tend towards being lean for most practical purposes, and often the more hydraulic they are, the less sand they can comfortably carry.

Although historic lean limes occasionally have strength values attributed to them, these can lie anywhere within a surprisingly wide range between the lowest strength of a pure lime to perhaps the middle of the feebly hydraulic range. However, these values are not so much a useful category in themselves but more a product of circumstance, and they have therefore not been included in Figure 1.

These historic lean limes were common materials, but largely existed due to sub-standard production, storage and transport. Lime was often inadequately burnt, leaving particles of limestone in the mix. In addition, a wide range of dirt and impurities could also be included, particularly if poor quality stone had been used at the outset. In addition, the burnt quicklime could partially air-slake during storage and transport. All these factors meant that a significant proportion of the lime might be inert, thus acting as aggregate and thereby reducing the amount of sand which the lime could subsequently carry.

These poor quality limes could also have partially hydraulic properties derived from some of these uncontrolled impurities, either from clay in the limestone, from included soil, or in cheap, poor quality coal used for burning. Any set resulting from their use would inevitably have been very variable, not just between sources, but between batches from the same supplier, and so could not have been relied on by the end users in any useful way.

Current practice

In recent years, the practice of specifying NHL3.5 hydraulic limes for general use in almost any circumstances has become widespread and is often even advocated by official agencies, sometimes without actually considering the circumstances of any particular building. This has the practical advantage of providing a material that is relatively easy to use and which has a conveniently quick initial set to give rapid frost and weather resistance. Before the mid-nineteenth century, however, limes of this strength would have been quite rare throughout much of the UK, and would normally have been reserved for civil engineering works including bridges, tunnels, docks and canals. They can be much stronger than necessary for masonry or render on historic buildings.

This frequent error is compounded by another misleading character-istic of BS EN 459-1:2010, that is, the establishment of the compressive strengths of limes at only 28 days old. This also comes from cement chem-istry, as an alite set will be substantially complete at this age. A belite set, however, will often continue gaining strength for 90 days and more. Thus all the NHL classes of hydraulic lime tend to produce finished mortars that are much stronger and harder and less permeable or flexible than is gener-ally recognized. The compressive strengths of the NHL limes are shown partially displaced in Figure 1 to indicate the possible increase that can occur between the 28-day classification point and the likely final strength – an increase which is not acknowledged in the standard.

In spite of the widespread acceptance of the lime revival, we are still therefore often using materials that are significantly harder than those originally used on historic buildings. The consequences of this lack of appreciation of the issue remain uncertain.

It is also worth noting that the three NHL classifications are drawn quite widely and it is theoretically possible for a single product to conform to all three grades at the same time. While it seems highly unlikely that any reputable manufacturer would try to take advantage of the latitude this offers for commercial gain, it does demonstrate that simply specifying a grade of NHL for critical conservation works may not be sufficiently precise. In such circumstances it is recommended that the manufacturers be contacted for the actual test results achieved by their products, as the practical behaviour of one brand of NHL3.5, for instance, can be very different from another, even though both might be manufactured to a similarly high quality.

Because it was impossible for many years to obtain feebly hydraulic limes, it became relatively common to try to obtain a comparable binder by adding NHL to a pure lime putty. This method can certainly work and has been widely used with success, particularly in Scotland and Scandi-navia. But this approach was called into question by English Heritage's moratorium some years ago on the use of hybrid mixes for grant-aided projects. It is crucial to recognize that under these circumstances the first proportion of hydraulic lime added will be inert. Thus a putty + NHL3.5 mix will give actual results closer to those which might be expected of a putty + NHL2 mix. In all cases, however, other variables can cause problems and site-specific trials are strongly recommended.

HISTORIC AND MODERN CEMENTS

As with the limes, there has historically been a wide range of different types of cements, characterized by varying proportions of alite, belite and free lime, and often also featuring significant proportions of aluminates. Another conspicuous gap in the spectrum of available limes and cements is that between NHL5 and modern OPC. This is particularly significant for the conservation of buildings from the later nineteenth and early twentieth centuries, including those from the early Modern movement. Although these often used cement as a critical component in their construction and appearance, the actual material available was very different from that found in builders' merchants today and few current equivalents exist.

An NHL5, perhaps with added pozzolan, can often be used to closely match the compressive strength of pre-First World War cements, although matching of other qualities, including colour and long-term behaviour, is not always achievable. Masonry cements, which tend to be propri-etary mixes of modern Portland cement and hydrated lime (usually with plasticizers and other additives), can occasionally be useful in this area. Their continued existence is a tacit admission of the relative unsuitability of OPC for laying brickwork or stonework. However, the appropriateness of a specific proprietary mix to an individual historic building cannot be guaranteed and there is little variety available.

A recent international EU-funded research programme, represented in the UK by the University of Bradford, has been studying the Roman cements widely produced during the nineteenth century, particularly in the UK from septaria occurring in London clay. These natural cements have a highly distinctive appearance, often with a strong reddish-brown colour which is notoriously difficult to match. The project's intention to re-commence manufacture of the material is quite an exciting develop-ment. Roman cements were, however, made effectively obsolete by the rise of early Portland cements after c. 1850, so this only partially fills the gap.

Some other natural cements have survived on the UK market, prin-cipally to suit the requirements of the potable water industry for low soluble chemical content and rapid set. The majority of these, however, are supplied as premixed mortars tailored specifically for the repair of water-retaining structures and are of limited use in conservation. That does not, of course, exclude the possibility that suitably formulated products may be developed in the future.

CONCLUSIONS

The question of 'lime versus cement?' is no longer an adequate model for the specification of mortars, renders or plasters in building conservation, but should instead be replaced by the question: 'which lime or cement?'. A suitable equivalent for most of the materials used historically, throughout all periods, is now available, but there are significant gaps in the overall spectrum which remain to be filled. This relatively wide availability puts the building conservation community in a position of strength which has never previously existed, although it must be recognized that each of these varied materials does need to be understood in its own right in order to achieve appropriate results. There are grounds to hope that a full spectrum of appropriate limes and cements for both conservation and new build may actually be available in the foreseeable future, although much work is still necessary.

This is a revised version of a technical overview that first appeared in 2006 in *Context*, the journal of the Institute of Historic Building Conservation.

Notes

1 Chief among these is perhaps the Dorking lime from the grey chalk of Surrey, which was used to build much of London from its Georgian squares to the West India Docks, as well as being exported far further away.

Bibliography

Bergoin, P., translated by Spano, U., 'Lime after Vicat', *Lime News*, Vol. 7, 1997. (Revised version in this volume.)

Holmes, S., 'Small scale lime production, hydraulic mortars, classification and standards', *Lime News*, Vol. 5, 1997.

Holmes, S., 'To wake a gentle giant – grey chalk limes test the standards', *Journal of the Building Limes Forum*, Vol. 13, 2006. (Revised version in this volume.)

Livesey, P., 'Portland cement properties through the ages', *Journal of the Building Limes Forum*, Vol. 10, 2003. (Revised version in this volume.)

Lime after Vicat

Pierre Bergoin

HISTORICAL BACKGROUND

There is a vast amount of evidence and documentation on the quest for better building limes prior to the advent of modern cement. From pre-Roman times to the beginning of the twentieth century, it was found that some limes were naturally strong and quick setting, while others were not. As the 'others' were the more common materials, the use of pozzolanic and other additives became standard practice to obtain the desirable setting and strength properties.

Some limes were found to be inherently very suitable for construction, as described in 1766 by Charles Fourcroy de Ramecourt:[1]

> The lime from Metz, with its quarries and production being the best in the North East [of France] is probably the best available anywhere up to now. It seems superior to the 'strong lime' from Piedmont and other Italian limes, and to the limes from Allais … A quantity of this lime, slaked in holes and covered with sand, has been found after a year as hard as stone … as found by the Romans using pozzolan and 'strong lime', by just mixing this lime with coarse river sand, it could resist a pick axe once hardened.

He would have probably made the same comments about limes from Aberthaw in south Wales, Charlestown in Fife, Scotland, and the Arden Lime Works at Darnley near Glasgow.

The eighteenth and nineteenth centuries saw the intervention of science, with initially sporadic, but later regular research into better building materials. Exchange of information through better communications improved the theoretical background of the industry, and the making of building mortars gradually moved from the craft of the artisan to industrial production, with the formulation of mortars becoming more specific to the particular works undertaken. Experiments were made on the various limes and mortar mixes, all of which appear to have been based on one underlying objective (conscious or otherwise), that is, to make mortars with early setting properties and rapid strength development.

Louis Vicat, in a period spanning over 35 years, drew from his predecessors and added his own considerable experience and acumen.[2] In 1818 he established in an irrefutable manner that 'the seizing of hydraulic mortars and their hardening were determined by the combination of lime with silica'.[3] Through his research and publications, Vicat did more than almost any other person to establish the scientific study of limes, cements and mortars in France, and he is justly regarded as the inventor of artificial cement. The Vicat company, founded by his son, is now one of the largest manufacturers of the material worldwide, and also produces 'Prompt' natural cement. The translation of his principal treatise into English by Captain J. T. Smith in 1837 did much to establish the cement industry in the English-speaking world.[4]

At the time of Vicat, lime production in France was widespread but its distribution was limited by inadequate infrastructure. Until the end of the nineteenth century, the lack of transport facilities required building materials to be produced locally, and sometimes lime kilns were positioned in the centre of the town – as was the case in Périgueux, the nearest town to St Astier, where the kilns were near the famous cathedral. For important work, materials were transported by river and waterways, but transport was in general limited to horse-drawn carts and rarely exceeded a distance of 50 km from the kiln. The advent of modern transport, especially railways, made transportation much easier, but until the end of the nineteenth century, there were still over a thousand kilns in France. Madeleine Fenart-Cuvelier notes that there were a minimum of five kilns in each commune at that time.[5]

Inevitably, the greatly improved infrastructure allowed large-scale producers of artificial cements to meet the industry's demand for maximum speed and strength. Cement became the primary commercial solution, mostly at the expense of hydraulic lime, the only other product capable of delivering anything like the properties sought. This hugely changed the way mortars were specified, mixed and used.

Traditional mortars are not easy to work with, demanding skill, dedication and understanding by the masons of the materials they were handling. Most of this skill was made redundant by the strength and the speed of setting of cement mortars. Traditional mortars also demand time to cure and attention after placing. Their use became too complicated and time-consuming, and therefore too expensive for the new needs of the building industry. To achieve hundreds or even thousands of years of durability was no longer a reasonable objective; the chief requirement was for speed and strength.

The increasing demand for cement, however, caused a consequent growth in the effort required to solve the problems created by its use. There is no doubt that the cement industry acknowledges many of the material's shortcomings and there are numerous textbooks and research reports on the factors causing the deterioration of cement, and many solutions put forward. None of these solutions, to the author's knowledge, will give cement mortars the enduring characteristics achieved by lime mortars.

If durable qualities are not required, should this be a concern? If it is more financially viable to replace one cement construction with another, in most cases bigger and more modern, is this a problem? Why maintain the arts and crafts of skilled specialists, at today's prices, if semi-skilled or even unskilled labour can do the job? Why build new constructions with materials that will give a better living or working environment if the client does not request it?

It would be unrealistic to deny the function that cement has, and will continue to have, in our ever developing world. However, the properties of traditional materials are recognized today as essential in the conservation and restoration of our historic buildings, while also presenting the new build sector with forgotten properties which have the ability to improve environmental standards.

VICAT AND HYDRAULIC LIMES

Vicat spent nearly 30 years establishing a survey of hydraulic limes and natural cements in France, starting in 1824. By 1845 he had travelled through the majority of the country, visiting 900 sites and listing the ones where good hydraulic limes were produced. Some of these were St Astier (Dordogne) and Le Tell (Ardèche) in 1833, Corbigny (Nièvre) in 1839, Roquefort la Bedoule (Bouches-du-Rhône) in 1840, Pamier (Indre et Loire) in 1844 and Antony in 1848. The following extract from the publisher's advert for his book about this survey, published in 1853, illustrates the comprehensive nature of his work:[6]

> One will find in this work, which comprises 76 départements, the compo-sition of calcareous materials, whether below or above the ground ...

naming producers able to supply materials suitable for hydraulic lime and
cement ... The impulse of this first work ... has produced ... a real revolution in the art of building and, if our important buildings of all kinds are
constructed and finished with a speed and economy previously unknown,
it is because the research of M. Vicat has discovered the country's wealth
of resources of which we had no notion.

As explained by Valentin Biston in a treatise on lime burning published
in 1836, Vicat proposed an initial definition and classification of hydraulic
limes based on what has been termed the 'Yield Theory':[7]

> There is a distinction between the common lime and the hydraulic lime.
> Common limes can be: fat, medium or lean and the term 'fat lime' is
> now adopted for quick limes which in contact with water will absorb in
> slaking between 2.6 and 3.6 parts of water to 1 part of lime. Medium limes
> will achieve a 2.3–2.6: ratio, and lean limes 1–2.3:1.
> Hydraulic limes are those that can set and harden under water in a few
> days without the help of any additive. These limes are suitable for under-
> water work and generally produce a good strength mortar when mixed
> with the necessary amount of water.
> Hydraulic limes are normally lean, often medium but never fat.
> However, lean limes are not always hydraulic as they could be the result of
> calcic lime badly burned (the 'biscuit' effect).

Biston also tried to classify hydraulic limes by colour: 'hydraulic limes
are sometimes white or with very little colour, more often they have a
light grey tint ... and other times yellowish'.[8] He concluded, however, that
'coloured limes are not necessarily hydraulic', adding that:[9,10]

> ... the name hydraulic lime should not mean however, that these limes
> give good results only when used in water. Their superiority over the
> ordinary limes is equally incontestable for all general building whether
> exposed to the air or in underground work ... and we can agree from
> well established experience that the resistance of a good hydraulic lime
> mortar used in these circumstances is as good as that of medium strength
> building stones ...

THE CONCEPT OF HYDRAULICITY

The importance of silica

In 1813 Hippolyte-Victor Collet Descotils, professor of chemistry at the École
des Mines de Paris, had noted that insoluble silica in calcareous materials
became soluble when this material was burnt at the necessary temperature.

He deduced that silica combines with lime during burning, and it was this process that gave hydraulic properties to some limes. Vicat based his map of hydraulic lime production sites in France on this principle.

In 1906, Edouard Candlot confirmed that 'in 1818 Vicat demonstrated that all calcareous materials contained a percentage of clay, finely dispersed in their mass, are able to give, after burning, hydraulic properties'.[11] Candlot explained the transformation of a calcareous rock when burnt at a certain temperature and set out the causes determining set and hardening. During burning, lime reacts with the clay and combines with silica to form calcium silicate. This salt, when hydrated by the addition of water, is the essential agent for consolidation. He also noted that the non-combined silica and the alumina would form an aluminium silicate which would also hydrate. The presence of alumina accelerated hardening but did not seem relevant to the production of a good quality hydraulic lime.[12]

As already perceived by Vicat, it seems that lime (CaO) and silica (SiO_2) are the essential elements for hydraulicity while 'other elements present in a deposit, being clays or not, although not necessarily nocuous, were not essential for the production of hydraulic lime'.[13] The term 'clay' is justified if it is interpreted as a material containing silica. Therefore, the ideal situation would be to work with materials containing mostly lime and silica with little or no presence of other elements. Quarries with these qualities were (and are) extremely rare, and therefore Vicat's findings were based on the most common composition of calcareous deposits, popularly referred to today as 'clay contaminated'.

The word 'clay' is used below to be consistent with the terminology adopted by Vicat and others but subject to the interpretation given above.

Initial attempts at classification

Depending on the quantity of clay present in a calcareous material, the resulting lime would have greater or lesser hydraulic properties. This principle allowed Vicat to classify limes as 'feebly and moderately hydraulic, hydraulic and eminently hydraulic'.[14] These classes were empirically defined using the clay content and the setting time, but later refined through the addition of the 'hydraulicity index' (discussed further below), defined as the ratio of silica and alumina to lime. Candlot summarizes the resulting classification system as follows:[15]

a) Feebly hydraulic limes are those where the hydraulicity index is between 0.10 and 0.16. The clay content in the calcareous material varies between 5.3% and 8.2%. These limes set in 16–30 days.

b) In moderately hydraulic limes, the clay content is between 8.2% and 14.8%, and the hydraulicity index is between 0.16 and 0.31. Setting takes place between 10 and 15 days.

c) In hydraulic limes, the clay content is between 14.8% and 19.1%, with an index of 0.32 to 0.42, and setting time is 5–10 days.

d) Eminently hydraulic lime is produced from calcareous materials with a clay content of 19.1–21.8%. The hydraulicity index is between 0.42 and 0.59, and setting occurs between 2 and 4 days.

Outside these clay contents, a slow-setting cement is obtained when the clay content is between 21.8% and 26.7% with a corresponding hydraulicity index of 0.50–0.65. When the clay content is above 26.7% and up to 40%, the hydraulicity index will be 0.65–1.28 and rapid setting cements are obtained.

However, Candlot also observes that:[16]

> ... this classification cannot be absolute because ... if a certain proportion of clay is necessary to achieve hydraulicity, its degree will often be subject to the manufacturing process and to the uniformity of raw materials ... as it is often experienced that ... in the same burning, most often than not, materials with different indices are mixed, and that the final product is a mixture of limes with different hydraulicities.

This continues to be a problem today when mixed calcareous/clay deposits are exploited, as it is nearly impossible to maintain consistency of components in the naturally occurring raw materials. Many producers are obliged to make additions such as pozzolans, gypsum, ashes and even white cement to level out the properties of their industrial output, usually to the detriment of qualities that, especially in conservation terms, should be present naturally.

Candlot continues:[17]

> ... in practice, for work of certain importance, one uses only categories b, c and d, although the latter is quite rare, as rare are limes with an index over 0.42.

He adds 'that under the name of eminently hydraulic limes there are products where the index is below 0.42 but setting takes place in less than 2 days'.[18]

This is one of the anomalies that even today puzzles researchers wishing to adhere to the hydraulicity and setting time theories of Vicat, and which has resulted in new approaches.

Vicat explored many other fields of lime production including investigating and proving the possibility of producing artificial hydraulic lime.

However, this avenue has been more or less abandoned today due to the generally poor qualities of the artificial hydraulic limes that have actually been produced.

Calculation of hydraulicity index

Candlot calculated the hydraulicity index of a number of the best-known limes of the day in accordance with the method suggested by Vicat. This was achieved by arriving at the clay:lime ratio as shown in the following example.

A lime sample contains 94.7% calcium carbonate ($CaCO_3$) and 5.3% clay. The molecular mass of calcium oxide (CaO) is 56 and the molecular mass of $CaCO_3$ is 100.

The lime (CaO) content of the sample is therefore:

$$94.7 \times \frac{56}{100} = 94.7 \times 0.56 = 53$$

And the clay–lime ratio is:

$$\frac{5.3}{53} = 0.10 \quad \text{(hydraulicity index)}$$

Table 1 shows a range of values obtained by Candlot using this method.

Table 1 Examples of hydraulicity indices from well-known French hydraulic limes.

Lime producer	SiO_2 Total	Al_2O_3	Fe_2O_3	CaO Total	MgO	SO_3.	Loss on Ignition	Hydraulic index
Molineaux	10.45	3.00	3.15	57.80	0.55	0.75	10.90	0.42
Marans	13.70	5.90	2.70	58.10	1.40	0.75	18.20	0.34
Le Teil	23.13	1.72	0.73	63.70	0.07	1.35	9.60	0.39
St Astier	21.85	1.35	2.85	62.25	1.05	0.50	10.15	0.37
Tournai	25.80	5.65	1.57	57.50	0.68	–	8.10	0.55

PROBLEMS WITH EARLY THEORIES OF CLASSIFICATION

Vicat's original classification of hydraulic limes and the various other early classification systems, which followed soon after, are not without their shortcomings. In general, they are empirically derived and based on some of the effects given by materials with different compositions but with no adequate basis on which to define a firm 'law' about hydraulicity. An examination of these various theories and classification systems produces the conclusions outlined below.

Classification related to setting time

This classification was based on the principle that limes with a setting time of more than one day were not hydraulic. Furthermore, the tests were related to pure lime paste and therefore not significant, as in most cases hydraulic limes are used in mortars and not as lime paste on its own.

The setting time of mortars depends not only on the hydraulicity of the lime used, but also on moisture content, meteorological conditions, the structure of the mortar, size and grading of aggregates, method of application, type of construction (e.g. bricks, stone, earth) and so on.

This attempt at classification is therefore not recommended as it severely distorts the actual factors upon which decisions about which mortar to use must be based.

Classification related to the cementation index

This classification should not be confused with the hydraulicity index discussed below. As indicated by Boynton, poorly founded empirical assumptions are again made to try and give hydraulicity a distinct entity, and the full range of relevant factors, as indicated above, is still missing.[19]

Moreover, this theory presupposes that combined SiO_2 is present as C_3S (alite), which is not the case in hydraulic lime, where the principal hydraulic component is usually C_2S (belite). If C_3S was present as the main hydraulic component, hydraulic limes could not be reworked as is normally possible.

Classification based on colour

Classifications based on colour were common in the early years of study but most were actually dismissed by their own authors.

Vicat's hydraulicity index classification

As demonstrated above, the Vicat classification based on the hydraulicity index is actually a speculative value, as it takes into consideration the result of a chemical reaction without analysing the reaction itself. This explains why anomalies can be found where, for instance, a lime with an index that classifies it as moderately hydraulic might behave, in its setting time, as an eminently hydraulic lime.

We know that limes are the result of the incomplete burning and subsequent slaking of calcareous material, and that when calcareous materials contain a proportion of silica in the raw state, this can combine with the free lime during burning to form calcium silicates. There is a direct quantitative relationship between the hydraulic and mechanical properties of hydraulic limes, and the resulting proportion of combined SiO_2.

Although the final mechanical properties are also affected by later carbonation of the calcium hydroxide ($Ca(OH)_2$), one way of making a basic classification of hydraulic limes easier and more realistic is to consider the actual amount of 'combined' silica, as only this can explain how it is possible to obtain different mechanical properties from the same lime with a constant amount of total SiO_2.

It is difficult to make sense of Vicat's formula for the hydraulicity index. The following observations can be made when considering, for example, the deposits of St Astier lime, which are more or less constant in terms of SiO_2 content and have a low content of Al_2O_3 and Fe_2O_3.

- The presence of actual 'clay' is not necessary to produce hydraulic lime, as already noted above. The notion that clays must be present has led to a specification for hydraulic lime which states that a 'clay content of $x\%$ is required', ignoring the fact that pure calcareous deposits infiltrated mainly by silica, although rarer, are actually more reliable in the production of hydraulic limes.
- More strikingly, if one follows Vicat's formula, a modern moderately hydraulic lime such as NHL3.5 might have a hydraulicity index of 0.37. This would entail a compressive strength of about 50 MPa!

Vicat's calculation for this example is as follows:

Rock contents $SiO_{2\ Total}$ = 13.8% of which, incidentally, 10% is available SiO_2 and 3% is insoluble quartz, which should not be part of the equation!

CaO_{Total} = 46.0% giving 82% $CaCO_3$

Al_2O_3 = 2.6%

Fe_2O_3 = 0.7%

MgO = 0.4%

$$\text{Hydraulicity index} \quad = \quad SiO_{2\ Total} \quad + \quad \frac{Al_2O_3}{CaO_{Total}} \quad + \quad Fe_2O_3$$

$$= \quad 13.8 \quad + \quad \frac{2.6}{46} \quad + \quad 0.7$$

$$= \quad 0.37$$

By applying the Voinovitch or Deloye methods, which are alternative modifications based on obtaining the amount of combined silica by calculating the percentage of SiO_2 in the raw materials and after burning, the following figures are achieved (assuming a proportion of $SiO_{2\ Total}$ of 19.7%):[20,21]

Voinovitch SiO_2 combined = 5.8%

Deloye SiO_2 combined = 10.8%

The question then arises: what should the quantity for $SiO_{2\ Total}$ be in the Vicat formula? Should it be 'total' silica (with or without the insoluble content) or 'combined' silica? If so, should the Voinovitch or Deloye method apply?

For example, applying the Deloye method gives:

$$\text{Hydraulicity index} \quad = \quad SiO_{2\ Deloye} \quad + \quad \frac{Al_2O_3}{CaO_{Total}} \quad + \quad Fe_2O_3$$

$$= \quad 10.8 \quad + \quad \frac{2.6}{46} \quad + \quad 0.7$$

$$= \quad 0.306 \ (20\% \text{ less than Vicat})$$

The choice of values for denominator is even more complicated. Measuring CaO_{Total} in the finished product gives a value between 57% and 60%

for NHL3.5. The proportion of 7% CO_2 found corresponds to a residual, unburnt proportion of $CaCO_3$, which will decrease the total CaO by approximately 9%. During slaking, there is a further decrease of CaO_{Total} of 9%. This would give a proportion of CaO combined with SiO_2 of:

$$60\% - (19\% + 9\%) = 32\%$$

To try to find a solution that will make the hydraulicity index work, a denominator that is CaO combined + CaO of hydration (which is 19%) could be used. The result would be:

$$\frac{SiO_{2\,Deloye}}{CaO\ combined + CaO\ from\ Ca(OH)_2} \quad or \quad \frac{10.8}{32+19} = 0.21$$

This presents yet another different value.

It is clearly time to look elsewhere for a reliable and constant method that identifies the hydraulic properties of materials. While Vicat provided the basis for all the research that followed his pioneering observations, it has to be acknowledged that his method results in values that are simply not correct.

The only occasion when Vicat's index is applicable is in cases where there is little or no $CaCO_3$ or $Ca(OH)_2$ present, as in the case of clinker. For example:

Typical cement clinker:

$$SiO_2 = 20.5\% \ / \ CaO = 65\% \ / \ Al_2O_3 = 5.3\% \ / \ Fe_2O_3 = 2.3\%$$

$$I = 0.42$$

This product will develop a compressive strength of approximately 55 MPa (N/mm^2) at 28 days, which is representative of the values actually obtained from modern artificial cements.

It would appear that hydraulicity is a function of the quantity of 'combined' silica and that this measure should therefore become an essential part of a modern classification of hydraulic limes. It is unfortunate that this method was not elaborated upon in the past, even when correctly perceived, as it was by Cowper in 1927.[22]

CLASSIFICATION ACCORDING TO EN 459-1:2010

European Standard EN 459-1:2010 sets out the classifications to be used for building limes within the European Union and manufacturers of limes are obliged to classify their products in conformity with this document and its companions within the EN 459 suite.

In classifying natural hydraulic limes, EN 459:2010 specifies the principal nomenclature to be in accordance with the materials' compressive strengths as tested at 28 days. But although the designations of NHL2, NHL3.5 and NHL5 appear to imply strengths of 2, 3.5 and 5 MPa (N/mm^2), the actual ranges of mechanical strengths specified as permissible for each grade are quite wide:

> Compressive strength @ 28 days using standard laboratory sands to EU Standards 1998 and subsequent:
>
> | | |
> |---|---|
> | **NHL2** | 2–7 N/mm^2 |
> | **NHL3.5** | 3.5–10 N/mm^2 |
> | **NHL5** | 5–15 N/mm^2 |
>
> Mortars made with normally available sands will typically show values of about 50% less.

Table 2 shows the new classifications given in EN 459-1:2010 compared with the old French nomenclature used prior to the first version of prEN 459 in 1998.

Artificial hydraulic limes (HL) are not regulated in their overall composition and disclosure of their contents is not required. Consequently, the presence of cement and other additives is almost certain, and these products are therefore normally unsuitable for conservation work.

The category of NHL-Z was introduced in 1998 to cover NHL limes with pozzolanic and other additives, reflecting the presence of such materials within the market. NHL-Zs were permitted to contain up to 20% of these additives, which could include cement in addition to other materials. These additives were often added to compensate for variation or poor quality in the natural material. In conjunction with further allowances made in the standards, the presence of sulfur trioxide (SO_3) could be between 3% and 7% (subject to a test of soundness), and the small amount of available free lime permitted in NHL products (3–8%) could also cause problems. In many instances these products could be acceptable in new build, but they were almost always unsuitable for restoration and conservation work. In recognition of the unsatisfactory nature of

Table 2 Classification of building limes before prEN 459-1:1998 and as in EN 459-1:2010.

OLD classification pre-1998	NEW classification 2010–	Main components
AIR LIMES	**Hardening only in contact with CO$_2$ in the air (carbonation)** **Classified according to their calcium oxide/magnesium oxide content as a percentage (90, 85, 80, 70)**	
Quicklime	Quicklime	Calcium oxide + magnesium oxide calcined and producing an exothermic reaction in contact with water.
CAEB	**CL** (calcium lime)	Calcium oxide or calcium hydroxide.
	DL (dolomitic lime)	Calcium and magnesium oxides (semi-hydrated) or calcium and magnesium hydroxides (hydrated).
	Hydrated lime	CL or DL resulting from controlled slaking of quicklime. Produced as powder, putty or slurry (milk of lime).
HYDRAULIC LIMES	**Setting and hardening in contact with water** **Carbonation also present** **Classified according to their compressive strength expressed in N/mm² measured @ 28 days**	
XHN 100 'Eminently hydraulic'	Natural hydraulic limes **NHL5**	Argillaceous or siliceous limestone burnt below 1250°C, slaked and ground to powder – no other additions allowed.
XHN 60 'Moderately hydraulic'	Natural hydraulic limes **NHL3.5**	
XHN 30 'Feebly hydraulic'	Natural hydraulic limes **NHL2**	
XHA Artificial hydraulic lime	Hydraulic limes **HL** 2, 3.5 and 5	Mixtures of calcium hydroxide, calcium silicates and calcium aluminates.
NHL-Z (1998–2010 only)	Formulated limes **FL** 2, 3.5 and 5	Limes with hydraulic properties consisting of CL and/or NHL with added hydraulic and/or pozzolanic material.

this classification, it was removed in the 2010 revision of EN 459 and is therefore no longer current.

The 2010 version of the Standard also introduced a new category of formulated limes (FL). These products are proprietary products formulated to achieve particular characteristics for specific uses. This category is not a direct replacement for the NHL-Z category, as it offers several significant advantages. In particular, these mixtures can contain a notably higher proportion of calcium lime (CL) and manufacturers are required to disclose the principal constituents. At this time all products classified as FLs have been designed for use in new building. In addition, they contain cement and/or other additives that make them unsuitable for conservation purposes.

Table 3 gives some typical relevant values contained in the EU and French standards, both compared to the values of St Astier NHL products. Other reputable companies are normally happy to make equivalent product information available on request.

The current European Standard is certainly welcome as it introduces a classification using a broad base of important parameters. In the case of natural hydraulic limes, however, the emphasis given to the compressive strength of the materials does not reflect accurately the real performance of NHL products, as the principle of testing compressive strength at 28 days is derived from the cement industry. However, studies prove that the true performance of NHLs is more accurately measured over a much longer timescale – normally at least several months. As a very general rule, for example, the compressive strength of an NHL mortar measured after one year can be as much as three times higher than the same mortar at 28 days. The crucial implication is that, if a specification is issued in conformity with the Standard strength measured at 28 days, it may well result in the actual production of mortars that could be far too strong at complete carbonation, particularly for conservation work. It would therefore clearly be preferable to re-classify NHL mortars according to their development of strength after one year, and then specify in accordance with the fabric of the structure for which they are intended.

Although this would be generally satisfactory for new build work, it is far more difficult to realistically calculate the mechanical performance of individual historic mortars in order to match them on the basis of compressive strength. In practice it is only really possible to guess this and then match the historic mortar with a modern equivalent by experience and 'feel'. As noted above, it is also misleading to attempt to match historic mortars by calculating their hydraulicity using the Vicat formula, or similar, as these cannot give correct indications either.

Table 3 Comparison of EN 459, French national standard and typical St Astier material figures.

NHL: PRINCIPAL CHARACTERISTICS
Mortars made with ISO 679 sand
Binder:sand ratio 1:1 by volume approximately (450g binder:0.83 litres sand)
Proctor's preparation, demoulding and curing conditions apply (EN 459-2)

	EN 459-1:2010			NPF 15.311			St Astier products		
	NHL2	NHL3.5	NHL5	NHL2	NHL3.5	NHL5	NHL2	NHL3.5	NHL5
Compressive strength at 28 days (N/mm^2)	2–7	3.5–10	5–15	2–5	3.5–10	5–15	≥ 2	≥ 3.5	≥ 5
SO$_3$ content (%)	3	3	3	3	3	3	0.65	0.45	0.55
Ca(OH)$_2$ content (%)	≥ 45	≥ 25	≥ 15	≥ 15	≥ 15	≥ 15	59	26	19.8
Expansion*	Maximum 2 mm			Maximum 2 mm			< 1	< 1	< 1

* According to EN 459-2:2010 section 5.3.2.2 or 'tablet test'. The method for assessing expansion included in section 5.3.2.1 of the standard, the 'Le Chatelier method', is not used in France any more as it is not considered appropriate for modern production because it allows expansion up to 20 mm.

If it is not possible to rely on Vicat or other methods, how can we be sure that we can safely find the degree of hydraulic behaviour of ancient mortars and produce satisfactory replacements? The only scientific method that is proving to be reliable in an increasing number of cases is to measure the proportion of combined silica.

CONCLUSIONS

After many years of study and examination of lime mortars from as early as 2000 years ago up to the nineteenth century, the following observations can be made.

1. Modern lime putties or hydrated powders are completely different from ancient lime mortars. Had today's lime putties been used instead of the ancient lime mortars, they and the structures built from them could not be analysed simply because they would not have survived. Modern lime putties are too pure and do not have enduring qualities. Ancient lime was obtained wherever possible from impure stone ('clay contaminated') or, when available, from pure limestone infiltrated with silica.

2. Impure or clay-contaminated stone contains silica and alumina, and often other materials that can provide hydraulicity. If these components are extracted using modern scientific methods, it can be seen that most, if not all, ancient mortars will have had some hydraulic properties. Vicat was correct in saying that silica, or silicates, are the reason for hydraulicity, but only now is it becoming clear that it is the soluble 'combined' silicates that are the real essence of this quality.

   Only in this way can we explain how it is possible to obtain products with different hydraulic set and strength from geological deposits with uniform proportions of raw materials (such as at St Astier). It is due to the different levels of combined silica in each class of product.

   If the level of combined silica in an ancient mortar is calculated, it is possible to obtain the correct parameter for making a modern equivalent. This means that for all old mortars showing a degree of hydraulicity, however large or small, we can be confident of providing a proper equivalent today. Obviously it is important to achieve this without using cement or cement-gauged mixes as their strength and undesirable chemistry may attack the very structures being conserved.

3. In the past, if the setting properties of a lime were not satisfactory, pozzolanic materials were introduced. Hydraulic properties were

always, therefore, required. However, it is fascinating to note from studies conducted since 1978 that only pozzolans with vitreous contents (silicates) can produce the necessary hydraulic effect.

A study of synthetic mixes containing obsidian and feldspar found that '... the former contained up to 65% of vitreous phase, the others from 20% to 0%'. The researchers further noted the reduction of pozzolanic properties in direct proportion with the vitreous phase in an alkaline environment.[23] Within the vitreous phase, the formation of calcium silicate hydrates (CSH_2) takes place in the form of a feebly crystalline gel, and one may also find tetracalcium aluminates (C_4AH_{13}) and possibly the presence of gehlenite and ettringite.

Therefore, when Smeaton noted that the pozzolan he used with blue lias lime in the mortar for the Eddystone lighthouse was producing a very slow set, it must have been a pozzolanic material with a relatively low vitreous content.[24]

Other researchers stated that pozzolanic activity is the result of:[25]

... the reaction of lime with the products of the alkaline attack on acid silicates ... this reaction creates hydrated components similar to the ones that are formed during hydration of clinker ...

In 1996, studies on the reactive fraction of volcanic pozzolans indicated that:[26]

calculations based up to today on the quantity of $SiO_{2\ Total}$ cannot be used to classify the reactivity of a pozzolanic material in a $Ca(OH)_2$ environment ... One has to look at the chemical composition of the *potentially soluble* [author's italics] vitreous materials as the only ones capable of reacting, instead of the total silica content.

It seems that also, with pozzolans, the 'combined' silica is the key to understanding hydraulicity. In practice, the use of the correct pozzolan with a feeble lime could create the same effect as obtained with natural hydraulic limes. However, the use of pozzolans can be potentially and, in some cases, economically unsuitable due to various factors. These include the sensitivity to water content in the mix, the variable setting characteristics, the granulometry and colour requirement, and the unnecessary complication of mixes and material costs.

If the introduction of pozzolans is intended to create a hydraulic effect, nothing would be of more benefit to our historic buildings than using a pure and natural hydraulic lime. Furthermore, if by matching the combined silica content of ancient mortars with the correct class of natural hydraulic lime available today, a relatively straightforward solution to the ever baffling question of how to analyse and reproduce these mortars would be achieved. This would allow conservators to restore and preserve our heritage, and put an end to the frequent use of speculative and potentially damaging mortar mixes.

This paper was translated and edited by Ugo Spano and was originally presented by Ugo Spano at a meeting of the Building Limes Forum in Newry, County Down, in 1998. It was subsequently published in *Lime News* in 1999. This is a revised version.

Notes

1 Fourcroy de Ramecourt, C. R., *L'Art du Chaufournier*, L'Académie Royale des Sciences & Arts de Metz, Metz, 1766.
2 Vicat, L. J., *Mortars and Cements*, 1828, translated in English by J. T. Smith, 1837. Facsimile edition, Donhead Publishing, Shaftesbury, 1997.
3 Boero, J., *Fabrication et emploi des chaux hydrauliques et des ciments*, Librairie Polytechnique Ch. Beranger, Paris, 1901.
4 Vicat, *op. cit.*
5 Fenart-Cuvelier, M., *L'age d'or des fours à chaux et cimenteries* [The golden age of lime kilns and cement], collection of old documents and postcards, 1991.
6 Vicat, L. J., *Recherches Statistiques sur les Substances Calcaires à Chaux Hydraulique et à Ciments Naturels*, Carilian-Goeury et Vor D'Almont, Paris, 1853.
7 Biston, V., *Manuel Théorique et Pratique du Chaufournier*, 2nd edition, Roret, Paris, 1836.
8 Ibid.
9 Biston, *op. cit.*
10 Biston, *op. cit.*
11 Candlot, E., *Ciments et Chaux Hydrauliques*, Ch. Béranger, Paris, 1906.
12 Ibid.
13 Vicat, 1837, *op. cit.*
14 Vicat, 1837, *op. cit.*
15 Candlot, *op. cit.*
16 Candlot, *op. cit.*
17 Candlot, *op. cit.*
18 Candlot, *op. cit.*
19 Boynton, R. S., *Chemistry and Technology of Lime and Limestone*, John Wiley & Sons, New York, 1980.
20 Voinovitch, I. A., et al., 'Analyse rapide des ciments', *Chimie Analytique*, Vol. 50, No. 6, 1968.
21 Deloye, F. X., 'Le calcul minéralogique: application aux monuments anciens,' *Bulletin de Liaison des Laboratoires des Ponts et Chaussées*, No. 175, 1991, pp. 59–65.

22 Cowper, A. D., *Lime and Lime Mortars*, BRE, London, 1927. Facsimile edition, Donhead Publishing, Shaftesbury, 1998.

23 Millet, J. M., Hommey, R. and Brivot, F., 'Dosage de la phase vitreuse dans les matériaux pouzzolaniques', *Bulletin de Liaison des Laboratoires des Ponts et Chaussées*, No. 92, 1977, pp. 101–4.

24 Smeaton, J., *A Narrative of the Building and a Description of the Construction of the Eddystone Lighthouse*, 2nd edition, G. Nicol, London, 1793.

25 Dron, R. and Brivot, F., Bases minéralogiques de sélection des pouzzolanes, *Bulletin de Liaison des Laboratoires Des Ponts et Chaussées*, No. 92, 1977, pp. 105–12.

26 Pichon, H., Gaudon, P., Benhassaine, A. and Eterradossi, O., 'Caractérisation et quantification de la fraction réactive dans les pouzzolanes volcaniques', *Bulletin de Liaison des Laboratoires Des Ponts et Chaussées*, No. 201, 1996, pp. 29–38.

Bibliography

Bernard, A., Millet, J., Hommey, R., and Poindefert, A., 'Influence de la température de cuisson et de la nature du calcaire sur la minéralogie des chaux vives', *Bulletin de Liaison des Laboratoires des Ponts et Chaussées*, No. 79, 1975, pp. 45–50.

Faure, B. and Geoffray, J. M., 'Les basaltes – planches expérimentales basalte-chaux', *Bulletin de Liaison des Laboratoires des Ponts et Chaussées*, No. 94, 1978, pp. 53–9.

Fournier, A., 'Les facteurs de qualité des chaux industrielles', *Bulletin de Liaison des Laboratoires des Ponts et Chaussées*, No. 79, 1975, pp. 73–7.

Fournier, M. and Geoffray, J. M., 'Les liant pouzzolanes chaux', *Bulletin de Liaison des Laboratoires des Ponts et Chaussées*, No. 93, 1978, pp. 70–8.

Geoffray, J. M. and Valladeau, R., 'Traitement des sables alluvionnaires par le liant pouzzolanes-chaux', *Bulletin de Liaison des Laboratoires des Ponts et Chaussées*, No. 93, 1978, pp. 86–91.

Gourdin, P., 'La cimenterie', *L'Usine nouvelle*, May 1984, pp. 146–55.

Lambert, P. and Rieu, R., 'La place des graves – pouzzolanes-chaux en technique routière', *Bulletin de Liaison des Laboratoires des Ponts et Chaussées*, No. 93, 1978, pp. 79–95.

Largent, R., 'Estimation de l'activité pouzzolanique – recherche d'un essai', *Bulletin de Liaison des Laboratoires des Ponts et Chaussées*, No. 93, 1978, pp. 61–5.

Millet, J., and Hommey, R., 'Etude minéralogique des pâtes pouzzolanes-chaux', *Bulletin de Liaison des Laboratoires des Ponts et Chaussées*, No. 74, 1974, pp. 59–63.

Millet, J., Fournier, A. and Sierra, R., 'Rôle des chaux industrielles dans leurs emplois avec les matériaux à caractère pouzzolanique', *Bulletin de Liaison des Laboratoires des Ponts et Chaussées*, No. 83, 1976, pp. 91–8.

Vicat, L. J, *Recherches Expérimentales sur les Chaux de Construction, les Bétons et les Mortiers Ordinaires*, Goujon, Paris, 1818.

Vicat, L. J, *Résumé des Connaissances Positives Actuelles sur les Qualités, le Choix et la Convenance Réciproque des Matériaux Propres à la Fabrication des Mortiers et Ciments Calcaires*, F. Didot, Paris, 1828.

Vicat, L. J., *Nouvelles Études sur les Pouzzolanes Artificielles comparées à la Pouzzolane d'Italie dans leur emploi en eau douce et en eau de mer*, Carilian-Goeury et Vor D'Almont, Paris, 1846.

Vicat, L. J., *Recherches sur les Causes Chimiques de la Destruction des Compose Hydraulique par l'Eau de Mer sur les Moyens d'apprécier leur Résistance à cette Action*, de Maisonville, Grenoble, 1858.

Structural Properties of Lime-Mortared Masonry

Pete Walker and Zhaoxia Zhou

INTRODUCTION

Until the twentieth century mortars for masonry construction were mainly lime-based: a mixture of calcium hydroxide, aggregate (sand or crushed stone) and water. Limes were typically produced in local kilns and slaked on site by the masons. Although the properties and quality of mortars varied accordingly, lime mortars were used successfully for thousands of years in a wide range of buildings and other structures.

Portland cement-based mortars largely replaced lime mortars after the Second World War. The benefits of cement over lime mortars included:

- improved consistency of performance;
- quicker strength gain, allowing faster construction;
- higher final strength, allowing thinner walls.

Cement mortars often include calcium hydroxide ('bag' lime) to improve workability, although plasticizers are also frequently used to replace lime altogether.

In contrast, lime mortars are weaker than cement-bound mortars, but this offers a different range of benefits:

- it allows the recycling of bricks as masonry units (not just as hardcore);
- they have higher vapour permeability, which is beneficial for the durability of masonry;
- they are more flexible, which is beneficial for accommodating movements arising from environmental or structural loadings.

Modern cement-based masonry buildings are typically characterized by movement joints placed every 6–12 m, but traditional lime-mortared buildings did not require these joints. This is principally due to the use of softer lime mortars, although it is important to note that traditional masonry walls are typically thicker than modern cement-mortared walls.

The ratio (by volume) of binder to aggregate for both lime and cement mortars is normally between 1:2 and 1:3. This mix ratio provides lime mortars with compressive strengths typically between 0.5 and 2.5 N/mm^2 at 91 days. Cement:lime:sand mortars are normally specified by the proportions of the constituents, such as: 1:¼:3 (strongest), 1:½:4½:, 1:1:6, and 1:2:9 (weakest). The final strengths achieved by lime mortars are generally comparable with those of the weakest cement mortar mixes at 28 days, while higher strength cement mortars can achieve strengths of more than 20 N/mm^2. However, the performance of mortars is always dependent on many factors other than just the proportion of binder to aggregate, including:

- the water to binder ratio;
- the characteristics of the aggregates;
- the environmental conditions during hardening;
- potential dewatering through laying with dry bricks;
- age.

The developing use of Portland cement mortars in the last century was accompanied by the development of British Standards for the structural design of masonry walls and other building elements. Early Standards, such as BS CP 111 from 1948, included structural design data on the performance of lime-mortared masonry.[1] However, with various revisions over time, structural design data for lime-mortared masonry was removed from the standards, reflecting the declining use of lime mortars in the UK. The latest standard for structural design masonry, EuroCode 6, was introduced in 2006 and, like its predecessor, BS 5628, includes no design data to support use of lime-mortared masonry.[2] EuroCode 6 uses performance-based limit state design: mortars are specified according to their physical performance attributes, including compressive strength, rather than the traditional prescribed mix design that relied on the specification of mix ratios between binder and aggregate as mentioned above. The move towards strength performance-based design is considered by many to be detrimental to the wider re-adoption of lime mortars as their strength is generally much lower than cement-based mortars.[3]

LIME MORTAR STRENGTH DEVELOPMENT

Strength is important to the use of mortars, as it significantly affects the overall load capacity of masonry. Although the low strengths of lime mortars ensure that any structural movement occurs along the joints between the masonry units, without unduly stressing the individual bricks or blocks, these low strengths also limit the capacity of masonry to resist structural loadings. However, the significance of this depends on the overall structural action. For example, a six times increase in mortar strength, from 2 to 12 N/mm², leads to only a 40% increase in the compressive strength of the resultant masonry.[4] According to provisions of EuroCode 6, the same change in mortar strength provides a tripling in shear strength and a 75% increase in flexural bond strength, although this also depends on the type of masonry units in use.

The strength properties of lime-based mortars depend on many factors. Material parameters include both the quantity and qualities of the binders, the aggregates and the amount of water added to the mix. Important environmental aspects include temperature and humidity during hardening, especially early in the setting process, as well as the rate of dewatering by the substrate on to which the mortar is laid (this is also often termed 'suction'). The quality of work, the elapsed time after mixing and before use, and the age of the samples at testing are also significant factors.

The compressive strength of mortars in laboratory testing is determined in accordance with BS EN 1015-11.[5] The compressive strength development of a 1:3:12 CEM II cement:lime:sand mortar is compared with a 1:2¼ (NHL3.5:sand) mortar in Figure 1. The same well-graded (Binnegar) sand was used in both mixes. The cement mortar mix reached 1 N/mm² at 28 days, matched by the hydraulic lime mortar after 91 days. Following the initial hardening by hydraulic set and drying out, further strength gain through carbonation is also evident in both mixes.

Sand grading also has an important influence on the compressive strength of lime mortars. The 28-day and 91-day compressive strengths for different mortars using different sands with an NHL3.5 lime are summarized below in Table 1. The mix proportions (1:2¼) were maintained constant. The use of a coarse-grained sand, instead of a fine-grained sand, effectively doubled mortar compressive strength at 91 days over the 28-day strengths. However, due to the possible effects of mortar dewatering on the overall properties of masonry, these increases may not necessarily be realized in actual wall strengths.

Table 1 Comparison of compressive strength gain over time in three NHL3.5:sand mortar mixes made with different grades of sand.

Aggregate	Compressive strength (N/mm²)	
	28 days	91 days
Coarse graded sand	1.13	1.52
Medium graded sand	0.63	1.20
Fine graded sand	0.50	0.76

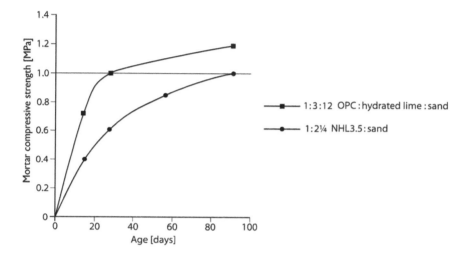

Figure 1 Comparison of compressive strength gain over time between 1:3:12 cement:lime:sand and 1:2¼ NHL3.5 hydraulic lime:sand mortars.

STRUCTURAL PROPERTIES OF LIME-MORTARED BRICKWORK

The capacity of masonry walls to resist gravitational and environmental loads is dependent on many, often interrelated, factors. These include:

- geometry (height; thickness);
- available restraints;
- nature of the loading (magnitude, position, duration, variation).

A key aspect is the strength of the materials. Three basic material properties are compressive, flexural (bond) and shear strength. Compressive strength governs the capacity to resist vertical, gravitational loadings.

Figure 2   Uniaxial compressive strength testing of a small brickwork wall in accordance with BS EN 1052-1:1999.

Shear and flexural strength govern the capacity of walls to resist lateral, wind loadings. Even non-loadbearing external walls must have sufficient flexural resistance to wind loadings.

In recent years two projects have studied structural performance of modern hydraulic lime-mortared brickwork. Results from an STI Link project led by the University of Bristol has formed basis for design guidance in the guide produced by the NHBC Foundation in 2008.[6] The University of Bath completed research on structural performance of NHL mortared brickwork between 2006 and 2009.[7] As masonry compressive strength is comparatively less sensitive to mortar strength, research has focused on flexural and shear strength.

Compressive strength

Uniaxial compressive strength of masonry is tested in accordance with BS EN 1052-1:1999.[8] One series of tests was completed at the University of Bath using NHL3.5 1:2¼ mortar (compressive strength 1 N/mm²) and extruded perforated clay bricks (nominal unit strength 55 N/mm²). Tests were conducted 91 days after construction. The average and characteristic compressive strengths were 12.1 N/mm² and 10.1 N/mm² respectively. (The

Figure 3   Testing the shear strength of a sample of masonry in accordance with BS EN 1052-3:2002.

characteristic strength is the value that 95% of all specimens of a given material would be expected to equal or exceed.)

The EuroCode 6 design value for these materials, assuming NHL lime mortar as a 'general mortar', is 9.1 N/mm². For comparison, CP111 (1948) would permit an allowable uniform compressive stress for the experimental materials of 1.4 N/mm². The STI Project also completed compressive strength testing, and the results have formed the basis of the NHBC Foundation guidelines.[9]

Shear strength

Test results for the initial shear strength of masonry, determined in accordance with BS EN 1052-3:2002, are shown in Table 2 below.[10] The characteristic values of initial shear strength range between 0.11 and 0.27 N/mm². The relationship between shear strength and normal stress was linear for the test specimens. The characteristic angle of friction (α), determined in accordance with BS EN 1052-3:2002 is also given below. The characteristic value of initial shear strength and friction angle for the 1:3:12 (cement:lime:sand) mortar at 28 days were similar to NHL mortar masonry at 91 days.

Table 2 Comparison of mortar compressive strengths and characteristic shear strengths to BS EN 1052, parts 1 & 3.

Mortar mix	Cement	NHL5		NHL3.5		NHL2
	1:3:12	1:2¼		1:2	1:2½	1:2¼
Age (days)	28			91		
Mortar compressive strength (N/mm²)	1.00	1.28	1.00	1.20	0.80	0.90
Characteristic shear strength of brickwork (N/mm²)	0.27	0.24	0.21	0.18	0.21	0.18
tan α	0.54	0.63	0.60	0.64	0.51	0.58

Flexural strength

Two methods are permitted to determine the flexural strength of masonry. The flexural strengths with planes of failure parallel and perpendicular to the bed joints can be determined in accordance with BS EN 1052-2:1999 using small wall prisms, as shown in Figure 4.[11] Alternatively flexural bond strength can be determined using the bond wrench test method in accordance with BS EN 1052-5:2005, as Figure 5.[12] Wall panel testing is frequently preferred by engineers, as it is the more established methodology in use in the UK. However, the more recently introduced bond wrench method requires significantly fewer materials and is generally easier to conduct. Comparative tests have shown good correlation between the two methods.[13] Flexural bond strengths for the brickwork obtained by the bond wrench method of testing are presented in Table 3 below.

The water absorption characteristics of the bricks significantly influenced the development of bond between the brick and NHL mortars. Both high and low absorption bricks developed poor bond. This is in contrast to the NHBC Foundation guide that shows reduction in bond strength only for bricks with water absorption above 12%.

Figure 4   Testing the flexural strength of a brickwork panel in accordance with BS EN 1052-2:1999.

Figure 5   Testing flexural bond strength using the bond wrench test method in accordance with BS EN 1052-5:2005.

Table 3 Flexural bond strengths for a range of brickwork types and samples obtained using the bond wrench test method in accordance with BS EN 1052-5:2005.

Brick type	Mortar	Age (days)	Average mortar strength (N/mm²)	Bond wrench strength			
				Mean (N/mm²)	Char. (N/mm²)	CV %	Range (N/mm²)
Berkeley Red Multi				0.46	0.36	12.2	0.36–0.52
Staffordshire Slate Blue Smooth			1.00	0.23	0.15	14.2	0.18–0.28
Hardwicke Welbeck Autumn Antique	1:2¼ (NHL3.5)	91		0.09	0.03	52.8	0.05–0.12
Chester Blend				0.33	0.17	32.9	0.15–0.54
Cheshire Weathered			0.93	0.35	0.12	33.6	0.21–0.49
Holbrook Smooth				0.49	0.28	31.7	0.38–0.68
Berkeley	1:3:12	28	1.00	0.35	0.30	8.1	0.30–0.40
Red Multi	1:2 (NHL3.5)		1.34	0.61	0.45	15.1	0.43–0.75
	1:2½ (NHL3.5)		0.78	0.38	0.31	11.1	0.30–0.47
	1:2¼ (NHL2)	91	0.90	0.29	0.21	18.0	0.22–0.39
	1:2¼ (NHL5)		1.28	0.63	0.45	16.3	0.40–0.80

Ibstock Ltd brand name	Staffordshire Slate Blue Smooth	Berkeley Red Multi	Cheshire Weathered	Holbrook Smooth	Chester Blend	Hardwicke Welbeck Autumn Antique
Water absorption (%)						
Average:	2.3	5.1	6.2	7.7	8.3	16.0
Coefficient of Variation:	14.9%	6.9%	12.3%	6.7%	6.0%	5.0%
Initial Rate of Water Absorption (kg/m².min)						
Average:	0.1	1.3	1.1	1.0	1.9	2.4
Coefficient of Variation:	35.0%	7.3%	13.7%	11.8%	10.8%	8.9%

CONCLUSIONS

Lime mortars gain strength at a slower rate than cement-based mortars. Carbonation is important to both the rate of strength gain and to the final strength. This slower rate of strength gain has consequences for both the testing properties of lime-mortared masonry and the early age performance of walls. Lime material performance should therefore be measured after 91 days rather than the 28 days commonly used for cement-based materials.

The lower strength and stiffness of lime mortars has benefits for movement detailing and the recycling of masonry units. However, lower material strength may have structural limitations that require thicker walls or greater restraints.

The flexural bond strength of hydraulic lime-mortared brickwork is controlled by the water absorption characteristics of the masonry units. The variation in strength performance ensures that design recommendations may prove conservative when compared against measured values.

Typical strengths of hydraulic lime-mortared brickwork are broadly comparable with low-strength cement-mortared brickwork. In the continued absence of robust design data, it is likely to remain necessary for structural engineers seeking to specify lime-mortared masonry in new works, or to evaluate properties of historic materials, to commission specific laboratory tests to determine actual performance levels. This is likely to remain a deterrent to the wider adoption of lime-mortared masonry in new construction until further tests have been completed that will improve understanding and confidence in material performance.

This paper was previously published in Volume 17 of the *Journal of the Building Limes Forum*, 2010. This is a revised version. Some of the data within this paper has previously appeared elsewhere.[14]

Notes

1 British Standards Institution (BSI), *BS CP 111. Structural recommendations for loadbearing walls*, BSI, London, 1948, 1964 and 1970.
2 British Standards Institution (BSI), *EN 1996-1. Eurocode 6. Design of masonry structures: General rules for reinforced and unreinforced masonry structures*, BSI, London, 2005.
3 Verhelst, F., Kjaer, E., Jaeger, W., Middendorf. B., van Balen, K. and Walker, P., 'Masonry – sustainable, contemporary and durable: Anachronism, bold statement or visionary outlook?', *Mauerwerk*, Vol. 15, No. 2, 2011, pp. 118–22.
4 British Standards Institution (BSI), *EN 1996-1*, 2005, *op. cit.*

5 British Standards Institution (BSI), *EN 1015-11. Methods of test for mortar for masonry: Determination of flexural and compressive strength of hardened mortar,* BSI, London, 1999.

6 Yates, T. and Ferguson, A., *NHBC Research Paper NF12: The use of lime-based mortars in new build,* NHBC Foundation, Milton Keynes, 2008.

7 Zhou, Z., Walker, P. and d'Ayala, D., 'Strength characteristics of hydraulic lime mortared brickwork,' *ICE Proceedings: Construction Materials,* Vol. 161, No. 4, November 2008, pp. 139–14.

8 British Standards Institution (BSI), *EN 1015-11,* 1999, *op. cit.*

9 Yates, *op. cit.*

10 British Standards Institution (BSI), *EN 1052-3. Methods of test for masonry: Determination of initial shear strength,* BSI, London, 2002.

11 British Standards Institution (BSI), *EN 1052-2. Methods of test for masonry: Determination of flexural strength,* BSI, London, 1999.

12 British Standards Institution (BSI), *EN 1052-5. Methods of test for masonry: Determination of bond strength by the bond wrench method,* BSI, London 2005.

13 *BRE Digest 360. Testing bond strength of masonry,* BRE, Garston, April 1991.

14 Zhou et al., *op. cit.*

Portland Cement Properties Through the Ages

Paul Livesey

HISTORIC CONTEXT

Since its invention in 1824, and subsequent patenting by Joseph Aspdin of Leeds, the development of Portland cement has been of clear interest to historians of construction. It is also, perhaps more importantly, a key to understanding the conservation of structures dating through nearly the past two centuries. A large proportion of the United Kingdom's building stock was built using the various versions of cement, and it is important to understand how these buildings work in order to predict their useful life, and to design for their conservation.

The early days of cement production were deliberately shrouded in mystery as cement manufacturers vied with one another to succeed in the marketplace. The early Portland cements had to struggle against the established Roman cements, and subterfuges such as carrying trays of brightly coloured minerals into the kiln were reported as ways of confusing or misleading rivals and competitors. Isaac Johnson is widely credited with the first real understanding of the principles involved, if not the first true production of Portland cement. Even he reports that he spent many fruitless months pursuing the possibility of calcium phosphate as the base following analysis of competitive cement from William Aspdin's production. Whether this was a successful subterfuge, or merely contamination between batches of cement production and use of the kiln to burn bones for the manufacture of fine china, is an interesting question.

'RELIABLE' RESEARCHES

We remain indebted to John Grant of the Metropolitan Board of Works in London for carrying out a systematic series of tests for strength, and for publishing these in 1866 and 1871 while engaged in building the main drainage system for the capital.[1] His testing regime was so well organized, and so much more reliable than those which had preceded him, that we can be confident of the comparative values for his tests. Less reliable, however, are attempts to correlate his tests with modern values from current tests, as although these are derived from the same basic principles, they differ significantly in detail.

For a measure of reliability in the ability to consider the difference between old and new test performance we can turn to the researches of Professor A. W. Skempton, as reported in his paper read at the Science Museum, London, on 5 December 1962.[2] He examines the work of Grant, adding and comparing the work of Bauschinger (1879) and Bohme (1883) in Germany. He further explores results from Michaelis in Germany where the basis for the 'modern' strength test by mortar prism was laid down

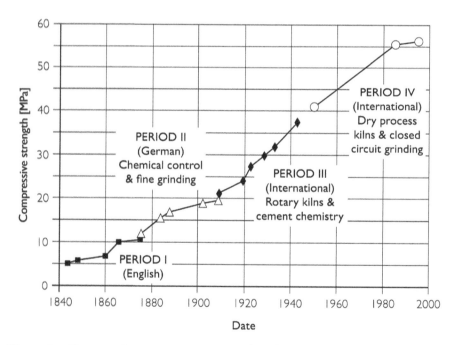

Figure 1   Comparative compressive strengths of Portland cements, 1840 to 2000 (Livesey, P., Early Cements, American Lime Conference, West Virginia, 2003).

in 1876. In his conclusions Skempton develops and carries forward the strength development 'curve' of cement production up to 1950. Reasonably comparable and reliable data to extend this curve to the present day can be drawn from articles by Corish and Jackson and a later update by Corish, both published in the journal *Concrete*.[3,4] Figure 1 shows the combined results of all these reports.

Even with the above comparators, however, the absolute strength values can only be at best indicative, as we now know the importance of sand properties, water content, curing conditions, and age on such strength tests. It is unfortunately impossible to numerically evaluate all of these retrospectively for the various series of tests carried out over this time span.

DEVELOPMENTS IN PORTLAND CEMENT

A comparison of the chemistry of the various Portland cements shows a slight theoretical increase in the cementation index over time, but even the raw materials of the early nineteenth century were actually capable of making a creditworthy modern Portland cement. The reason this was not achieved was due to poor understanding and crude production techniques.

The first major breakthrough in the improvement of the production process came in Germany in the 1870s with the development of proper chemical control, allowing both a better final product and greater consistency. Thereafter, the main changes related to the physical aspects of the process, allowing raw materials with increasing lime contents to be consistently burned at temperatures high enough to activate the silicates.

From the original 'beehive' kilns, operating a batch process, the technique moved to continuous burning, but at similar temperatures. Preparation of raw materials by wet processing dominated production in England and France, where chalk was the main source of calcium carbonate, initially requiring massive settling ponds, or 'backs', to prepare the mineral slurry before feeding it into the kilns. A useful development of this was to utilize waste kiln heat in 'chamber' kilns to pre dry the chalk slurry feed. In contrast, in hard rock areas, shaft kilns using dry materials were more efficient, allowing continuous operation and consistently higher temperatures.

The most significant technological breakthrough came with the introduction of the rotary kiln. Initial trials by Ransome and Stoke in England lacked the necessary financial backing to succeed, and it was left to Seaman and Hurry in the United States to find the final solutions which could achieve the temperatures and efficiency to unleash the massive twentieth-century cement industry. Since then, the further refinements of preheater kilns, and improvements in grinding technology, have moved the industry forward in efficiency of production and consistency of quality, but with only relatively small increases in the typical compressive strengths of the final product.

PORTLAND CEMENT CHEMISTRY

It is the chemistry, or rather the mineralogy, of Portland cement that determines its characteristics and performance. While the basic raw material mix of limestone and clay has changed little, the way they are treated, their fineness, the temperature and the length of time to which they are heated, the cooling process, and the final additions during grinding, all play their part.

The principal chemical reaction in Portland cement is the interaction of calcium oxide, produced by calcination of calcium carbonate, and silica released when clay minerals decompose. The early Portland cement of 1824, as with Frost's 'British cement' of 1822, would inevitably have had little calcium oxide–silica interaction because the temperature in the kiln was too low to activate the silica. It was not until William Aspdin, at Northfleet, and Isaac Johnson, at Swanscombe, developed larger kilns capable of higher temperatures, that significant amounts of dicalcium silicate (belite) came to be produced, together with small amounts of low-reactivity tricalcium silicate (alite).[5]

The increasing understanding of the importance of the silicates, and of the correct proportioning of the raw materials, brought the step change to what we would now recognize as the superior hydraulic limes of the later nineteenth century. These, however, still lacked the consistency of temperature and time in burning to produce reactive alite for true Portland cement.

Figure 2 illustrates the temperature gradient required to produce the necessary mineral changes: it was only with the development of the rotary kiln at the beginning of the twentieth century that consistent high-temperature burning became feasible on an industrial scale. This opened the way to steadily increasing alite proportion, and the significant gains in product strength which were consequently achieved.

Table 1 shows the trends in product composition which took place over the twentieth century after the widespread introduction of the rotary kiln. The nature of the hardened product changed as overall alite proportion

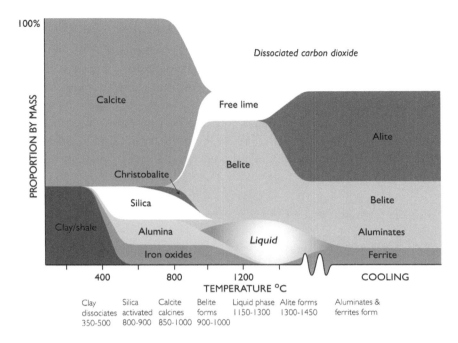

Figure 2   Portland cement – phase changes with temperature (Livesey, P., Early Cements, American Lime Conference, West Virginia, 2003).

increased: the greater proportion of reactive silicates led to a more dense hydration product, with longer chain silicate molecules and lower amounts of interstitial calcium hydroxide (portlandite). The composition trends shown here are somewhat simplified, as the nature of alite, in particular, changes with process parameters that affect its crystal size, impurities and hence its reactivity.

Table 1   Trends in Portland cement composition over the twentieth century.

	Typical proportions of principal active ingredients (%)		
	Alite (C_3S)	**Belite (C_2S)**	**Free lime (CaO)**
1914–1922	35	20	8
1928–1930	40	30	6
1944	40	30	5
1960	45	25	2.5
1980	55	20	2
1990	55	20	1.5
2005	60	20	1.5

COMPARATIVE CURRENT MATERIALS

In considering which current materials might compare with Portland cements used over the past two centuries the simplest parameter is that of compressive strength. However, we have seen that chemistry also plays a major role. The use of these comparators has also to be tempered with the knowledge that mortar, render and concrete performance relies on much more than density and strength. For these products to meet the durability performance there needs to be a balance of strength and chemical resistance requirements.

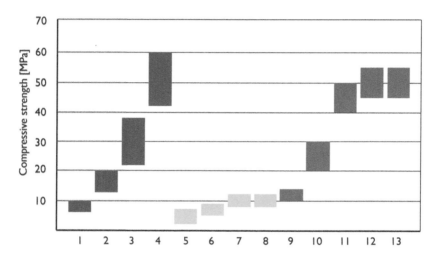

Key:

1 Portland cements of Period I (1840–70)
2 Portland cements of Period II (1870–1910)
3 Portland cements of Period III (1910–40)
4 Portland cements of Period IV (1940–2000)
5 Natural hydraulic limes Class 2
6 Natural hydraulic limes Class 3.5
7 Natural hydraulic limes Class 5
8 Natural hydraulic lime Class 3.5 with 10–30 % by volume of metakaolin or microsilica
9 Portland cement:CL90 combinations (65:35 to 50:50 by weight or 1:1 to 1:2 by volume)
10 Masonry cement
11 Portland fly ash cement CEM II/B-V (80:20 to 65:35 PC:Pfa by weight)
12 Portland limestone cement CEM II/B-LL (90:10 to 80:20 PC:limestone by weight)
13 Portland blastfurnace cement CEM III (50:50 to 30:70 PC:blastfurnace slag by weight)

Figure 3 Comparative standard compressive strengths of Portland cements and a range of other binders currently in use.

The materials used in the Foresight lime project have been considered with cements currently on the UK market.[6] They include natural hydraulic limes in all three grades (2, 3.5 and 5), masonry cements, Portland cements, Portland composite cements (i.e. cements with Portland cement clinker as the base combined with secondary materials) and combinations of limes with pozzolans. The comparative standard compressive strengths of these are shown in Figure 3. It should be noted that the standard strength is that of a 1:3 by weight cement:sand mortar with a water–cement (binder) ratio of 0.5. Typical masonry mortars mixed to give adequate site workability will have higher water content, and resulting compressive strengths approximately half the standard strengths shown here.

Figure 4 compares the strengths and water absorption characteristics as measured by capillary absorption. It can be seen that even small proportions of Portland cement as in 1:1:6 and 1:2:9 cement–lime mixes result in a densification of the matrix and reduced absorption compared with NHL mixes of equivalent strength, justifying the need to consider wider issues than just the quoted compressive strength of a particular binder.

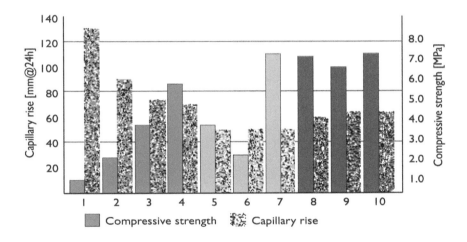

Key:

1 NHL 2 @ 1:3 vol/vol;
2 NHL 3.5 @ 1:3 vol/vol;
3 NHL 5 @ 1:3 vol/vol;
4 NHL 3.5 + 20% MK @ 1:3 vol/vol;
5 PC/CL90 @ 1:1:6 vol/vol;
6 PC/CL90 @ 1:2:9 vol/vol;
7 MC @ 1:4 vol/vol;
8 CEM II/B-V @ 1:6 vol/vol;
9 CEM II/A-LL @ 1:6 vol/vol;
10 CEM III/B @ 1:6 vol/vol.

Figure 4 Comparative properties of capillarity and compressive strength for typical site workability mortars proportioned by volume.

Table 2   Development of compressive strength over curing time for typical NHL and Portland cement-based mortars.

	Typical values of 2005 materials (MPa) – 1:3 mortar prisms; w/b 0.5			
	28 days	91 days	1 year	2 years
NHL2	4.0	6.8	10.2	11.8
NHL3.5	5.5	9.3	13.9	16.1
NHL5	7.0	10.7	15.0	17.2
1:1:6 PC/CL90/sand*	5.0	5.7	6.4	6.8
CEM I PC	60.0	65.0	67.0	68.0

* w/b 1.1

Finally, it should be noted that compressive strengths are typically measured at 28 days curing time. This is appropriate to the higher reactivity and faster strength gain of alite, so is generally representative of the actual performance of Portland cement-based mortars. However, the slower reactivity of belite means that final strengths of NHL-based mortars are achieved well after this time has elapsed. Table 2 shows the comparative compressive strength gain characteristics of NHL and Portland cement-based mortars up to a curing time of two years. Note that, as with mortars tested above, the majority of these had a water:binder ratio of 0.5, so mortars mixed for site use will normally achieve significantly lower strengths.

This paper was first published in Volume 10 of the *Journal of the Building Limes Forum* in 2003. This is a revised version.

Notes

1   Grant, J., *Experiments on the Strength of Cement*, F. & F. N. Spon, London, 1875.
2   Skempton, Professor A. W., *Portland Cements, 1843–1887*, excerpt from the Transactions of the Newcomen Society, Vol. XXXV, 1962–1963.
3   Corish, A. and Jackson, P., 'Portland cement properties – past and present', *Concrete*, July 1982.
4   Corish, A., 'Portland cement properties – updated', *Concrete*, January/February 1994.
5   Blezard, R., 'Chapter 1. History of calcareous cements', *Lea's Chemistry of Cement and Concrete*, Fourth Edition, ed. Hewlett, P. C., Arnold, London, 1998.
6   Allen, G., Allen, J., Elton, N., Farey, M., Holmes, S., Livesey, P., and Radonjic, M., *Hydraulic Lime Mortar for Stone, Brick and Block Masonry*, Donhead Publishing Ltd, Shaftesbury, 2003.

Developments in the Use of Lime in London c. 1760–1840

Ian Brocklebank

INTRODUCTION

The period 1760 to 1840 is of particular interest in the study of lime use in London. It was essentially during this period that available mortars, renders and plasters were transformed from those that had been known by craftsmen for centuries to the precursors of many of the modern materials now used throughout the general building industry.

This historical period covers most of the early industrial revolution, in this case from John Smeaton's experiments into hydraulic lime mortars for the Eddystone Lighthouse[1] (1756–1759) to the beginning of the great railway building boom of the 1840s, which coincided with the first commercial production of modern types of cement. After this, materials manufactured in bulk began to be transported around the country in large quantities, eventually resulting in the almost complete replacement of craft-based locally derived materials with proprietary industrially produced supplies.

The transformation in materials that occurred resulted from an intense period of speculative scientific investigations, changes in the methods of material production and revised geographical sourcing of materials for political reasons. The transformation is both well documented and clearly visible through its effects on the changing architecture and townscape of the city.

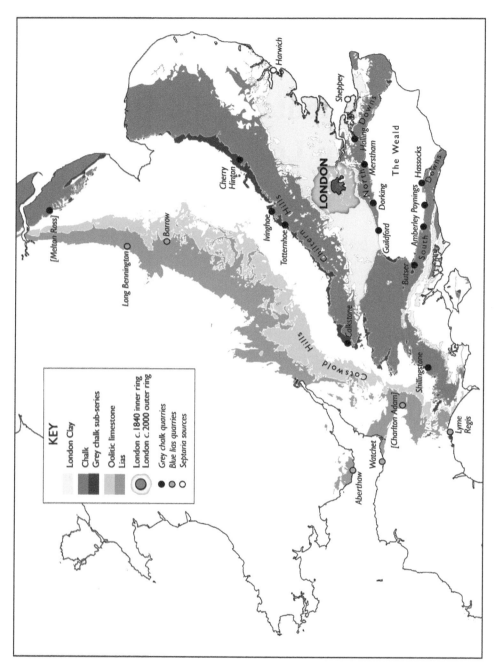

Figure 1 Principal geology and lime sources in south-eastern England *c.* 1840. Based on information from the British Geological Survey and Stafford Holmes.[2]

GEOLOGY

Figure 1 shows the principal relevant geological strata in south-eastern England and their geographical relationship with London. The city itself sits mostly on the distinctive London clay from which the yellow-grey London Stock bricks were made. This clay is geologically very recent, dating from the Eocene epoch; it is overlaid in places with even younger river gravels (omitted for clarity from Figure 1), which have been used from Roman times as a good source of building sands and aggregates. The clay lies within a dip, or syncline, in the underlying Cretaceous chalk beds, which continue below the city from the Chiltern Hills to the North Downs, and from which much of London's fresh water is extracted. The Weald to the south is an eroded anticline where the same chalk strata originally continued up and over in a dome shape between the North and South Downs.

To the north-west of the Chilterns, as one moves further away from London, successively older rocks are exposed, each dipping down to the south-east below the later strata above. Significant among these are the Jurassic oolitic limestones, which produce perhaps the finest building stones in England in a distinctive geographical belt from Dorset, through Bath and the Cotswold Hills to Oxford, and then via the east Midlands to Lincoln and the North Yorkshire Moors. Underlying the oolitic lime-stones are the Lias beds from the early Jurassic period, much of which is a mixture of limestones, clays, siltstones and some sandstones, usually found in distinctive thin layers.

All these geological formations are most easily accessed from their scarp faces, where the beds have been eroded in depth to naturally expose their lower strata. Because of the orientation of the beds, these faces are generally found on the north-west side, except in the Weald, where they are the inward-facing slopes of the North and South Downs. This acces-sibility means that these are the locations where the greatest number of quarries were historically found.

TRADITIONAL LIMES IN LONDON C. 1760

For centuries up to and including the middle Georgian period, and prob-ably for almost as long as the city has existed, two principal types of lime had been in use in London – most often referred to by the building trades as 'chalk lime' and 'stone lime'. Both these materials were bought by builders as quicklime for slaking in the yard or on-site prior to use, either directly to form mortar in a 'hot mix' with the sand (and a moderate amount of added water) or with excess water to form a putty. Hydraulic works were usually carried out with mortars based on imported trass.

Chalk lime

As its name indicates, chalk lime was derived from the upper layers of the chalk strata and was therefore a very pure material (perhaps 95–98% pure calcium carbonate), with no hydraulic properties. As can be seen from Figure 1, these upper layers of the chalk are widespread and easily accessible throughout much of southern England and this was therefore a common, inexpensive and easily available material.

It is important to recognize that chalk lime was always seen primarily as a material for internal works, principally plastering, though it was also used for limewashing. For both uses it would normally have been run to a putty and allowed to mature. As a building lime it was regarded as significantly inferior because of its lack of any hydraulic set and its use for external or structural works, such as brickwork, would always have been regarded as poor or cheap workmanship. Assumptions in recent years (or at least in the early years of the lime revival) that pure, chalk lime putty was normally used for bricklaying or general building work are almost entirely erroneous.

Finest quality chalk lime putty, also frequently called 'fat lime', would have been very similar to a modern matured pure lime putty and was the material of choice for the best interior plastering, providing both the workability and the precision needed for high quality running and modelling. Wherever possible, it was thoroughly matured and sieved to ensure absolutely complete slaking and to remove any grit, inert powder or other impurities. This was therefore a rare and relatively expensive material, normally produced by plasterers solely for their own use, as their professional reputations depended at least partly on its quality.

Also available on the London market was 'lean lime', which was normally a poor quality variant of a chalk lime, containing many unburnt and overburnt particles, and usually a range of various other impurities from both the original extracted material and the firing process. This was essentially a poor quality option, although added impurities could give it a very weak and unpredictable hydraulic set. It was, however, a persistent feature of lower quality building works, despite many attempts to discourage its use and can often be seen in plain historical plasterwork of the period, especially in the interiors of lower status buildings.

Stone lime

Stone lime was also produced from chalk, but in this case from the lower strata, geologically known as the grey chalk sub-series. This grey chalk is less pure than the upper strata and contains a useful proportion of clay (typically up to perhaps 12%), which provides a mild hydraulic set. Stafford Holmes discusses this type of lime in considerable detail (see page 199), noting that its importance has not been fully appreciated during the lime revival to date and that it is unfortunately not properly recognized in the current British and European lime standards.

The common historic name for this material was adopted by lime suppliers specifically to distinguish this slightly hydraulic lime from the non-hydraulic chalk lime. The term was not actually correct but was brought in as an early form of marketing, invoking the simple and wide-spread (but incorrect) belief, derived from Vitruvius, that a harder stone would give a harder lime after burning. This long-standing assumption was accepted and substantially unchallenged until the publication of John Smeaton's initial researches, where the role of the clay content in providing the hydraulic set was first fully recognized.[3]

Although the grey chalk beds underlie the pure chalk throughout southern England, the depth and continuity of the younger strata make them substantially inaccessible except where the erosion of a scarp slope has taken place. The grey chalk quarries were therefore consistently located on the narrow exposed band at this scarp (Figure 2). In many of these quarries (notably those at Amberley, Dorking, Halling and Butser, but probably most others), large quantities of chalk lime were also produced, not least because this enabled economic value to be obtained from the deep overburden that had to be removed to reach the more valuable grey chalk below. It is significant that the quarries of the only current UK producer of hydraulic limes, at Melton Ross in Lincolnshire, are also found in an equivalent geological location.

Limes from Dorking, Halling and Merstham were very widely used in London, partly because they were generally of consistent quality but also because they were the closest centres of production. Lime from Halling and nearby quarries had a particular advantage, as the River Medway here not only cuts completely through the North Downs from the Weald to the Thames estuary, thus naturally exposing the lower grey chalk strata, but also providing convenient and cost-effective transport by sailing barge directly into the city.

Stone lime was correctly used for all external work and for structural

Figure 2   Panoramic view across the Medway gap in the North Downs looking west towards Halling. Although now substantially overgrown, the sheer scale of the historic chalk workings is clear. On the skyline the North Downs dip down from the scarp slope on the south (left) towards the north (right) and the Thames estuary.

elements such as loadbearing external walls, works below ground level, external render and other general building works (Figure 3). Typically it was slaked by mixing quicklime 1:3 into damp sand, enough water being added to make a stiff mix, before being thrown through a coarse screen, banked up, and left to finish slaking for several days prior to actual use.

It was common practice on a London building site to make the mortar on a Friday for use during the following week. Immediately before use the mortar was reworked to restore its plasticity, preferably without adding any more water (although this may in practice have actually depended on the particular circumstances). Gerard Lynch discusses in detail the

Figure 3   Schomberg House, Pall Mall, London, 1698. Fine flat gauged brick arch made with putty joints: the putty used would almost certainly have been run from stone lime, rather than the pure lime putty with which we are now most familiar.

difference between 1:3 mixes made with quicklime or with putty (see page 219). Unlike mortars made with a putty, the addition of water to quicklime significantly increases its volume, resulting in a finished mix closer to 1:1.5 or 1:2 by volume, and this apparent richness is common in historic mortars.[4]

Although stone lime has hydraulic properties, the slowness of its set and its high fat lime content provides considerable benefits in workability and many craftsmen are as happy using this material as a pure putty-based mortar. If required, stone lime could be run to a putty by plasterers who needed particular durability as well as the traditional putty workability (see below) or by bricklayers for fine gauge work, in which case it would remain usable for perhaps two weeks. After this, its hydraulic set would begin to take and it would revert essentially to a conventional non-hydraulic putty in behaviour, the hydraulic component becoming no more than a very fine aggregate. In normal building works, the hydraulic action, although slow, was critical to obtaining quality, ensuring that the mortar would always take a set to the full depth of the work and could generally withstand all weather conditions. As noted by Major Percy Smith, 'a good sample should sensibly resist the fingernail at a month old'.[5]

Trass

Some stone limes from a few particularly clay-rich beds, such as the Sussex Clunch from near Lewes, had a sufficiently active hydraulic component to enable them to achieve a useful set under water, but the majority were fairly ineffective when submerged. For durable hydraulic works where mortars would be fully or partially submerged or regularly saturated (such as docks, bridges, canals, drainage works and the like), the use of mortars based on trass (also historically referred to as tarras or terras) had by the late eighteenth century been standard practice in south-east England for several centuries.

Figure 4
Zwanenburgwal, Amsterdam (photograph by Frank R. Yerbury, 1924). This type of construction has always been heavily reliant on trass-based mortars.

Trass is a volcanic earth which has been extracted since Roman times from the Eifel mountains (part of the Rhenish Massif) in north-western Germany. The traditional area of production in the Eifel was the Brohl Valley; the material is still available though it now comes from nearby Andernach. Both locations allow convenient export by water up and down the River Rhine and its tributaries.

The principal characteristics of trass are very similar to those of the much more famous volcanic earth from Pozzuoli near Naples; it has similarly strong pozzolanic qualities and therefore similar potential uses. When added to the free lime component in either a pure or hydraulic lime mortar, this pozzolan has the ability to give a much stronger and quicker hydraulic set to the end material. As well as acting as a pozzolan, trass assists the workability of dense mortars, helps protect the lime from attack by acids and modifies the performance of the lime to reduce water movement by capillary action.

'Trass' itself is actually a Dutch word, reputedly derived from the

Figure 5   Pre-formulated 'Trasskalk' (trass–lime) binder in use for historic brickwork repairs and re-rendering at the Nederlands Openluchtmuseum, Arnhem, summer 2003.

Italian word 'terrazzo', and there is a long and well-developed history of its use in the low countries for the construction of canals, water engineering structures, bridges and general brickwork, including solid ground floors and paving (Figure 4). Not only was the material easily obtained and cost-effective, but it provided durable and reliable hydraulic mortars for the distinctive high quality brick construction so prevalent throughout the largely low and damp terrain of the Netherlands and northern Flanders.

The importation of trass from the Netherlands for use as an addition to local lime mortars in hydraulic work, normally at a ratio of 1:2 of trass to lime, is recorded in the London region as early as the sixteenth century. By the mid-eighteenth century the importation of trass-based mortars as preformulated binders, ready mixed with lime, was well established. The material came in a range of four strengths, apparently as widely used in the Netherlands, although known in England in decreasing order of strength as 'Strong Trass', 'Strong Bastard Trass', 'Bastard Trass' and 'Slack Bastard Trass'. Interestingly, it is reported that only materials from adjacent strength classes could be mixed, as those further removed on the scale would not combine and set. The use of trass-based mortars within England at this time was considered a specialist skill, not normally found within the general building trades.[6] Figure 5 shows the use of trass today in the Netherlands.

STIMULI FOR CHANGE

This pattern of material use had been used and developed sporadically for centuries, becoming effectively universal during the extensive rebuilding and extending of the city of London in the century after the Great Fire of 1666. However, not all was well and several important factors were encouraging change. By 1760 these were combining to become urgent.

Problems with quality

The following quotation from a craft textbook of the period by R. W. Dearn leaves little doubt that there were significant and widespread problems with the quality of lime delivered to building sites in the capital:[7]

> In London, lime is for the most part little better than powdered chalk; which may be accounted for, in the first place, the chalk is seldom sufficiently burnt, and in the next, it is too long exposed to the air after it is burnt.

Although both chalk and stone limes were in theory useful products when properly understood by tradesmen (which they mainly were), the reality on-site was very different with most tradesmen becoming frustrated and disappointed.

It is appealing to imagine an industry based around skilful small-scale craft production of lime using local materials for local buildings, but in fact the widely disparate sources meant that quality was extremely variable and good material was notoriously hard to find. Despite the huge market for lime in the building of what was becoming the world's largest city and the capital of a growing international empire, the business of lime-burning remained very competitive and the records show that bankruptcies were not uncommon.

Fuel for lime burning was not only expensive but it was also in short supply, often being the most significant cost facing lime producers. Inevitably, there was a persistent temptation to overeconomize, and partially and underburnt material was commonplace. Although it was normal practice for unburnt lumps to be screened out after slaking, in some cases the ratio of unburnt to burnt material could be so poor as to make an individual mortar mix unusable.

The selection of material for burning was also not always what it might have been. As the parent chalk contains natural variability and scientific testing at production sites was not possible, it would not have been easy to predict the actual degree of hydraulicity that any particular batch of stone lime would achieve. Under these circumstances any producer who could maintain consistency of behaviour would have been able to charge for a premium product. However, such consistency and the valuable reputation that went with it could dramatically disappear as quarrying moved on to a slightly different strata or formation in the parent chalk.

Even if the material itself could be produced to the highest standards, its transport from kiln to site as quicklime, which took advantage of the significantly lower weight of the material in this form, left it open to air slaking and partial carbonation, particularly if the packing and protection was poor, the journey delayed or the weather less than ideally dry. The worst instances could deliver virtually inert material to site, under which circumstances Dearn's comments above would have been an accurate description.

Given that all these problems could so easily be combined, not only was the actual performance of the lime on-site open to question but, in the absence of any definitive scientific understanding of lime chemistry, it would be impossible to determine at which stage of the process the problem had occurred and thus who was to blame. Human nature being what it is, there can be little doubt that many unscrupulous members of the supply chain would have taken full advantage of the doubt and confusion surrounding the whole process. On balance it is almost surprising that any good quality material made it to market at all, although a moderate, if unpredictable, proportion apparently did.

London Building Acts

Ever since the Great Fire of 1666, the issue of controlling fire spread within a tightly packed city had been a major preoccupation of the authorities. Before 1666, most buildings in London had been timber, with masonry and brickwork reserved for high status and institutional structures. The fire had such a devastating effect that building in timber was banned from 1667 and it is from this date that the construction of London as a distinctly brick city began.

Legislation, however, had to be progressively tightened to reduce the risk of fire spread more effectively, not least with the London Building Act of 1774, which further restricted the use of timber on the exteriors of buildings. Earlier local building acts had introduced a range of detailed provisions, but this Act consolidated and extended these to cover the whole of the rapidly expanding urban area. Its comprehensive require-ments had a significant effect on the permitted styles of architecture

Figure 6   No. 46 The Butts, Brentford.   **Figure 7**   No. 40 The Butts, Brentford.

within the greater city; Figures 6 and 7 show two nearby buildings at The Butts, Brentford, which illustrate these changes well. Although Brentford was at the time a village well upstream from London, the legislation was adopted locally some years later.

No. 46 The Butts (Figure 6) is representative of urban domestic buildings constructed in the late seventeenth and very early eighteenth centuries. The structure of the house is of brick and the roof tiled, but there is still a fair amount of exposed timber on the façade, particularly the sash boxes to the windows, the detailing under the projecting eaves (actually banned in the City of London and Westminster in 1707 and 1709 respectively), and the door canopy and surround. This is essentially an example of what later came to be known as the Queen Anne style and predates fully developed Georgian domestic architecture.

No. 40 The Butts (Figure 7) shows the typical characteristics of buildings constructed throughout greater London, particularly after the introduction of the 1774 Act. Here, there is a solid brick parapet above the roof line to protect the roof and any dormer windows behind, and to remove the vulnerable eaves; the sash boxes to the windows are recessed behind four inches of brickwork and the door surround is no longer of timber.

The later style of building is conspicuous for its plainness and for the lack of articulation or ornament, often leaving architects seeking to create elegance or beauty little to play with other than the proportions of the window and door openings, and perhaps the occasional string course. The proportions of this particular building are perhaps not the most elegant, but the principle is clear and the exercise of pure proportion in the composition of façades came to be a matter of considerable sophistication and social status during the eighteenth century.

However, there remained a strong desire to ornament buildings and incombustible materials that would enable this at a more modest cost than the use of stone were greatly sought after. One example of the consequence of this was the rise of Coade Stone (actually a fine earthenware) for external details. There was also a strong impetus to the search for types of lime or cement that could take fine detail and yet endure without protection, unlike earlier 'stucco' made from pure and mildly hydraulic limes.

Political factors

During this time the import of trass from Europe for hydraulic mortar became difficult, expensive and dangerous due to political troubles. England and France were frequently at war between 1688 and 1815 (the Second Hundred Years' War), and conflicts with America (1775–83) and the Netherlands (1780–84) also disrupted cross-channel trade, as did the Belgian war of independence from the Netherlands (1830–31). The wars also accelerated the need for hydraulic construction works for military and naval installations. Research in this field therefore became extremely urgent and was encouraged by the Government, at least partly through the Royal Society.

EXPERIMENTATION AND RESEARCH

By 1760 the problems were so pressing and the potential rewards so great that many people from a wide range of backgrounds joined the race to find a solution. The number of patents awarded simply for new types of lime binders or 'cements' during the period under consideration (Table 1) is an indication of the activity within the field. Many of these materials inevitably disappeared from sight fairly rapidly, but several proprietary products and certain particular researchers' names lasted well and remain widely recognized.

Table 1   Patents awarded for modified binders and cements, 1760–1840.

1765:	Warks	1818:	St Leger
1772:	Rawlinson	1820:	Ranger
1774:	Liardet	1822:	Frost ('British Cement')
1776:	Adam brothers	1824:	Aspdin ('Portland Cement')
1779:	Higgins	1830:	Anderson
1796:	Parker ('Roman Cement')	1832:	Troughton ('Metallic Cement')
1809:	Earles	1834:	Martin
1811:	Dobbs	1838:	Greenwood and Keene
1811:	Frost		('Keene's Cement')
1816:	Atkinson ('Yorkshire Cement')	1840:	Francis ('Medina Cement')
1818:	Vicat & John		

James Parker's 'Roman Cement' is well-known and discussed further below, while Joseph Aspdin's 'Portland Cement' is acknowledged as the precursor of the Portland cement which forms the mainstay of the modern building industry (although the actual material is much changed). Vicat founded the cement industry in France and Keene's cement (actually gypsum-based) remained available until the 1970s.[8] Research continued beyond 1840 with a wide range of further new proprietary materials being introduced, though these are outside the scope of this paper.

Bryan Higgins

Dr Bryan Higgins MD delivered a well-attended series of public courses in chemistry in London in 1774, which formed the basis for his *Experiments and Observations made with the View of Improving the Art of Composing and Applying Calcareous Cements, and of Preparing Quick-Lime* ..., published shortly after he patented his 'cheap and durable cement'.[9] Higgins' understanding of the processes involved in the burning and preparation of lime were based on phlogiston theory and give an interesting insight into the state of scientific knowledge at the time.

Phlogiston theory was an early attempt to explain combustion and oxidation processes. It asserted that all combustible materials contained an 'element' called phlogiston which was driven off into the air by the combustion process, leaving the parent material in its natural state.

Higgins' application of this approach to lime production led him to believe that the difference between chalk and stone limes was simply a matter of the relative completeness of the burning process, rather than of their chemical composition. He assumed that a fully 'dephlogisticated' lime would both slake and set better, and that the purer the lime, the better it would be:[10]

> When first I noticed the quantity of chalk-lime which slakes latest or not at all, I suspected that that this difference might in some degree be owing to the admixture of argillaceous [clay] or other matter, but … I was convinced that the only impediment to their slaking consisted in their not being sufficiently burnt.

We now know this to be completely incorrect and his discarded suspicion to be actually the right answer. It is nevertheless interesting to see an intelligent and educated man striving to advance scientific understanding in the face of established myths. This was mainstream science in its day, and was still being authoritatively cited in Dearn's *Bricklayer's Guide* … some years later.[11] Phlogiston theory was fully overturned in the scientific community by Antoine Lavoisier in 1783, but its use in the understanding of lime tended to persist, at least informally, until the publications by Smeaton, Vicat and Pasley.[12,13,14]

John Smeaton

John Smeaton is justly recognized as the founder of civil engineering in Great Britain, although he originally trained in London as a maker of scientific instruments. His first and most famous civil engineering work was the building of the Eddystone lighthouse outside Plymouth harbour, constructed during the period 1756–59.[15] The year before the design and construction of the lighthouse began, Smeaton went on a study tour to the low countries, travelling primarily by canal, during which he studied the design of water engineering, drainage and canal structures, as well as both wind and water mills. Inevitably during this tour he came across the use of trass-based mortars for hydraulic engineering works and the insights he gained seem to have provided a useful reference for much of his subsequent career as a designer of harbours, bridges, canals and lowland drainage schemes.

> After this I went to see a Terras mill near Rotterdam … The Terrass stone … is a natural production and comes down to Holland from mines in Germany where it is dug in the manner of Coals, and the only Art in preparing it for use is that of reducing it to powder.[16]

At Ostend he visited sluice construction works then under way at the harbour:[17]

> All the brick and stone work being laid in Terras, I observed many hands employed in beating & preparing the mortar. They were divided into 2 companys, one being planted at each End of the Works, each company being under a shed covered with Boards at Top, but open at the sides; they beat the terras in a sort of tubs set in the ground, with wooden pestils [pestles] slung in a Cord to the end of a pole spring; there seemed at least 100 men in each company.

However, the construction period of the Eddystone lighthouse coincided with the first three years of the Seven Years' War (1756–63) and it seems probable that the importation of Dutch trass mortars would have been extremely difficult, if not impossible. In addition, Smeaton was also concerned that trass mortars had a tendency to exhibit crystalline growth forward from the joints over time, which caused difficulties with navigation and lock operation. He discusses this problem in both *A Narrative of the Building and a Description of the Construction of the Eddystone Lighthouse*[18] and in a letter of 1775 (see quote below taken from a compendium of his reports published after his death in 1792). Sir Charles Pasley, however, suggested in 1838 that this issue of trass expanding forward from the joints was not usual and resulted from incorrect usage in parts of Britain, and that it was not normally apparent in the Netherlands.[19]

There was no pozzolan with the performance to match trass (crystalline problems notwithstanding) in the British Isles. However, the political situation did allow the importation of a material from the original site near Pozzuoli in Italy, which Smeaton considered to have more predictable long-term behaviour. To find the most suitable British lime to match this rather than accept the prevailing myths of the day, he carried out a series of simple laboratory tests and thorough comparative mortar trials based on a wide range of domestically sourced limes, firmly establishing in the process the importance of the proportion of clay in the parent material.[20]

> The best kind of lime for water-works that I know of, is from Watchet, in Somersetshire, Aberthaw, in South Wales, and Barrow, in Leicestershire; and the strongest composition I know, is made by an equal quantity of lime, striked measure in the dry powder, after being slaked and sifted, and of pozzelana, ground and prepared ..., and, if put together with as little water as may be, and beaten till it comes to a tough consistence, like paste, it then may be immediately used; but if suffered to set, and it be afterwards beaten up a second time to a considerable degree of toughness as before, using a little moisture, if necessary, it will set harder, but not so quick.

Table 2 John Smeaton's order of preference for limes for hydraulic work.[21] Geological type added by the author. Please refer to Figure 1 for quarry locations.

No.	Name/Location	% clay	Geological type
1	Aberthaw, South Wales	13.0	Blue lias
2	Watchet, Somerset	12.0	Blue lias
3	Barrow, Leicestershire	21.4	Blue lias
4	Long Bennington, Lincolnshire	13.6	Blue lias
5	Sussex Clunch, Lewes (Nr Hassocks)	18.7	Grey chalk
6	Dorking, Surrey	5.9	Grey chalk
7	Berryton Grey Lime, Hampshire (Butser)	8.3	Grey chalk
8	Guildford, Surrey	10.5	Grey chalk

This composition is of excellent use in jointing the stones that form the lodgment for the heels of dock-gates and sluices, with their thresholds, &c. when of stone.

A second kind of mortar is made by using the same proportion of ingredients as terras mortar, that is, two measures of lime to one of pozzelana beaten up in the same manner, and which, if used with common lime, will fully answer for the faces of walls either stone or brick that are exposed to water, either continually, or subject to be wet and dry; in which last case the pozzelana greatly exceeds the terras, as also in its lying quiet in the joints as the trowel has left them, without growing as terras does. As a piece of economy, I have found that if the mortar last mentioned is beaten up with a quantity of good sharp sand, it nowise impairs its durability, and increases the quantity. The quantity of sand to be added depends upon the quality of the lime, ... thus, if the lime is of such quality as to take two measures of sand to one of lime, then one measure of pozzelana and three measures of sand will satisfy two measures of lime ...

The first composition will assuredly acquire the hardness of stone under water, and in twelve months will be as hard as Portland.

The hardening of the second and third depends greatly upon the quality of the lime ... yet there is scarcely any lime with which the materials, well beaten up, in the proportion specified, will not acquire a very competent degree of hardness under water.

Table 2 gives Smeaton's stated order of preference for limes for hydraulic work. The publication of Smeaton's research inevitably led to a wave of considerable interest in the use of blue lias lime in London, as it promised many of the characteristics so eagerly sought after and particularly the possibility of consistent quality and long-term durability in weathering environments. Although it was quickly recognized as an excellent material for stucco, blue lias lime did not achieve immediate widespread use, partly because Smeaton's work took some time to reach publication and partly because it remained, like trass, essentially an expensive specialist material requiring rather different skills and understanding from the craftsmen using it.

James Parker

Another likely reason for the slow take-up of blue lias limes in the capital was the almost instant success of Parker's 'Roman Cement', which was patented in 1796 and came into full production in 1798. Parker was a clergyman who seems to have discovered the benefits of calcining septarian nodules from the London clay almost by accident; the result being a strong naturally hydraulic lime with a distinctive dark brown colour. The material was not burnt at a high enough temperature to become a true cement in the modern term, but the speed of its set (as little as fifteen minutes) and the durability of the finished material were unprecedented.

Parker set up his manufacturing plant at Northfleet in Kent on the south side of the River Thames just on the edge of the greater London area. This location allowed the septaria to be taken from Sheppey upriver by barge to the works, where it was burnt using coal as a fuel and ground to powder before being shipped out to building sites across the capital.

The original patent lasted until 1810, by which time the success of the material had led to many imitators, with other types of Roman cement being produced elsewhere from the London clay, including the Isle of Wight ('Medina Cement') and Yorkshire ('Whitby Cement').

As these sources of stone were all somewhat different, the term 'Roman Cement' actually covers a wide variety of similar materials. The type produced from septaria dredged from the seabed near Harwich, in particular, came to dominate the London market during the first half of the nineteenth century, not least because winning the stone needed in bulk was much easier and cheaper here than excavating the cliffs at Sheppey (Figure 8), although one proprietary version ('English Cement') apparently mixed both sources.

Roman cement enabled significant changes in the prevailing architectural styles within London in the first part of the nineteenth century. While there were a range of attempts to develop the standard Georgian brick building types by rendering them externally with proprietary oil-based mastics, these almost invariably failed and the prevailing architectural

Figure 8   Sea cliffs on the Isle of Sheppey from where James Parker obtained septaria from the London clay. Nodules still naturally erode out here and can be readily found at the base of the clay: two can be seen here broken open (though not to be confused with the weathered concrete blocks from the failed local sea defences mixed among them).

style of the Regency era, typically including a full coat of articulated and painted stucco over the entire elevation, was primarily enabled throughout the country by the durability of Roman cement stucco. Also popular at this time was the early 'Gothick' revival, which in many cases also used Roman cement for its relative cheapness and durability, and its ability to be worked into, and retain, extremely intricate detailing.

Parker's cement was greatly appreciated for being noticeably harder and faster-setting than blue lias lime, although the colour was not popular. The finished stucco could be painted in oil paints without harm, partly to disguise its colour, but as a result allowing the use of a durable, visually homogeneous and thoroughly impermeable external skin for the first time in architectural history. As we now know, this brought with it new problems of its own, but it marks a significant change in the history of both general building and architecture, and might reasonably be regarded as the true beginning of the modern period in both. Figures 9 and 10 show examples of the use of Roman cement.

Figure 9 Gunnersbury Park House, West London, 1806, by Alexander Copeland. Rendered overall and finished with oil paints. Quoin detailing and run mouldings also of Roman Cement, as is the cast detailing to the porte-cochere balustrade. The detail shows the distinctive dark brown colour of the Roman Cement where the paint has been lost.

Parker's Northfleet works were later bought by Joseph Aspdin's son William, and used from 1845 to produce the first commercial 'Portland Cement' burnt at a high enough temperature to count as a fully modern material, and which itself effectively replaced Roman cement in the London building trades.

Figure 10   Hadlow Tower, near Tonbridge, Kent, 1838, by George Taylor. Interestingly, the base coats of the lower parts are in lime, overlaid with Roman cement, a mixture which has failed comprehensively. High level detailing is built up with mould-cast and bench-run Roman cement elements set into an overall Roman cement render, run over and bulked out with brick slips.

HISTORICAL SNAPSHOT IN 1840

Thomas Leverton Donaldson's article on 'Stucco' for the *Encyclopaedia Metropolitana* published in 1840 offers a fascinating insight into the production and use of lime and early cement mortars and renders within London at the time.[22] Donaldson himself was an experienced practising architect, the first Professor of Architecture at University College London and a founder of the Royal Institute of British Architects. Particularly when read in conjunction with Sir Charles Pasley's *Observations on Limes* of 1838, a very clear overall understanding of the state of play at this crucial historical time can be gained.[23] Pasley was no less distinguished than Donaldson, becoming eventually a Major-General of the Royal Engineers, responsible for many years for teaching all young engineer officers how to build the military structures necessary throughout the Empire of the time.

Grey chalk limes

As noted above and discussed by Holmes, the vast majority of mortars used in London had for centuries been the 'feebly' hydraulic limes from the lower grey chalk beds of the North Downs.[24] Donaldson describes these geological formations and their product thus:[25]

> The bottom bed of the great deposit of chalk is considerably thicker than the upper ones, and contains no flints; in colour it is a less pure white or grey, so that many parts of it make a tolerably good building stone. In composition it is not uniform, the proportion of slightly ferruginous clay that it contains notably increasing from the top to the bottom of the bed; the lower parts moulder by exposure to the weather, and the lowest of all not only moulder, but are more or less slaty in structure; that is in the state of true marl. That part of the grey chalk which is used for water-cement contains various proportions of clay, from six or eight up to twenty-five percent, and after burning has a pale brownish-yellow colour. It is known in the London market by the name of Dorking lime, there being very extensive quarries of it near that town, as well as at Merstham and Halling.

However, although the grey chalk still provided the raw material for the bulk of the capital's mortars, it was by 1840 no longer normal practice to burn the material at the quarry and transport it all the way to site as quicklime. In preference, larger suppliers had grown up who purchased the chalk from the quarry and transported it raw for burning much nearer the city.

Lime production in London

Lime production at this time was becoming quite heavily industrialized in character. There were now many kilns within London itself, generally on riverside wharves to enable convenient and cost-effective transport. Burning close to site rather than at the quarry had the advantage of greatly reducing the decay of the product through air slaking and carbonation during transport. Fuel for burning was generally coal from Newcastle, for which a huge coastal trade had grown up to serve the capital's vast and steadily increasing appetite for it.

A significant proportion of this lime burning was already an integral part of a wider industrial complex, as can be seen from the kiln designs in Figure 11. The kilns in question are both running kilns, giving continuous production and are integrated with coke ovens. The coal was burned to coke in the surrounding smaller ovens, the volatile gases given off being used to fire the lime kiln. The coke produced was sold to blacksmiths and other industrial trades for various types of metalworking and manufacture. This form of production also gave large quantities of extremely good quality lime (from carefully selected raw material), thoroughly burnt, containing very few impurities, and giving rise to almost no waste ash by-product.

The first kilns of this type were actually built to 'Mr Hearthorn's patent' near Maidstone in Kent in 1825. Perhaps the largest range of combined kilns was at Walker's works on the Greenwich peninsula where a row of eight large kilns, each with four attached coke ovens, maintained continuous operation. In time the production of lime was scaled down as a result of the rise of cement. However, the need for coke remained as long as horses needed shoeing and the volatile gases given off from the coke ovens became 'town gas' or 'coal gas', piped out across the city to light buildings and streets in the latter half of the nineteenth century. Walker's old lime works at Greenwich eventually became the largest gas works serving London until being made redundant by the arrival of natural gas from the North Sea in the 1960s. The derelict gasworks site is now the location of the Millennium Dome.

Limes in use

Donaldson gives a range of interesting notes which illustrate the general usage of limes within London at this time:[26]

> A cubic yard measure filled an inch above the top is taken by London builders to be equivalent for a hundred of lime.

(a)

(b)

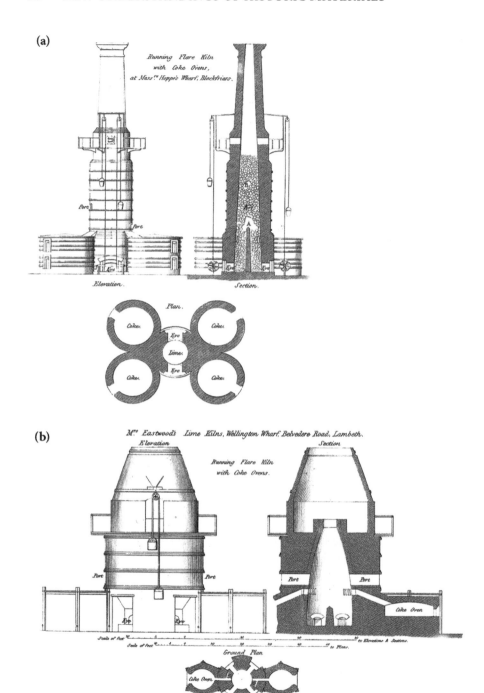

Figure 11   Examples of lime kilns combined with coke ovens from (a) Blackfriars and (b) Lambeth. (Courtesy of the RIBA Library.)

Lime is paid for by the quantity delivered. If lime is carried any distance in a wagon or cart on a rough road it will settle by up to a full seventh.

Lime is delivered lump or ground. Lump is better for chalk lime as the core is preserved [presumably against air slaking and carbonation] ... Ground is often preferred by builders using greystone lime, as it mixes better.

He recommends the following method of slaking lime to minimize excess water:[27]

Walnut-sized pieces are put in a basket and immersed until the surface of the water begins to boil. The basket is then withdrawn and allowed to drip a little, then the lime turned out into boxes. Cover with straw and keep in a dry place. This method is often used by plasterers when they cannot find chalk limes as it prevents a strong stone lime from blistering.

Pasley attributes the origin of this method of slaking by immersion to a French publication by de la Faye dated 1777 and notes that it was common-place in France. Although a rather ad-hoc method, it produced perhaps the first hydrated lime powder, similar in many ways to modern materials, which could be more easily stored and transported in bags or casks than either a putty or quicklime.[28]

For aggregates:[29]

There are beds of rivers, such as the Thames, which produce a sand admi-rably calculated for all purposes of construction.

Limes are considered rich or poor depending on the amount of sand they can accommodate, rich limes admitting a greater proportion. Thus chalk lime should not have more than two parts of sand to one of lime. Dorking, Merstham or Halling lime will take three parts of sand to one of lime.

(Note that this refers to quicklime proportion.[30])

Architects often specify two parts of sand to one of lime, but it is best to give lime as much sand as it will bear; too little is as bad as too much.

Various mixes for special works are included:[31]

A good blue pointing mortar is made from air slaked and sieved greystone lime. Mix three parts of sand to two of lime and add six pounds of lamp-black to seven or eight hods, and one part of smith's ashes.

For a red pointing mortar substitute a mixture of Spanish brown and yellow ochre for the lamp black.

For pargeting flues take one yard of cow dung and three barrows of strong lime and mix with plenty of water.

Concrete

Concrete deserves particular mention as its use for foundations had been pioneered in 1817 by Sir Robert Smirke for the footings of the new penitentiary at Millbank in Pimlico, after the previous architect's first two (of six) radial segments of the building had failed.[32] The results were so successful that he went on to use the same methods at the new British Museum (constructed 1825–50). Apparently this 'admirable compound of lime and gravel' provided a 'solid bed in indifferent soils below stone and brickwork'.[33]

> Ballast is mixed with dry lime [presumably blue lias lime, see below] and then wetted. It is wheeled in barrows and thrown down from a height of at least three yards. It is then puddled and treaded down by men specially employed. The lime slakes in place, and its expansion in a confined space is invaluable for assisting consolidation. Where speed is required it should be mixed with hot water where it will set immediately.

Blue lias lime

In 1840, blue lias lime had only just been introduced to the London market and had been used in just a few places. Donaldson notes that it had been very successfully used for the stucco façades of Belgrave Square (designed in the 1820s by George Basevi for the developer Thomas Cubitt) and for pointing the paving on the west side of St Katharine's dock (completed in 1828 by the engineer John Rennie).[34]

> Of hydraulic mortars for subaqueous construction blue lias is the most valuable. It is produced for the London market from Lyme Regis in Dorset. [It] … is less used in the metropolis than it deserves, being about twenty-five percent dearer than Dorking lime.
>
> The stone is burnt near Gravesend. It requires less fuel, but care is required in burning to avoid vitrification. Mr Watson, the engineer at Ramsgate, burns it in domed kilns as … [Figure 12]. He throws a few buckets of water on the heap of calcined lime and covers it over with a thin coat of sand. After a day or two it is put through a pugging mill to be ground into mortar.
>
> Mr Gladdish slakes conventionally, but it takes longer: eighteen or twenty-four hours to fall to a fine powder … too much water and it will set hard. It is ground very finely and left to stand for some weeks for plasterers use, otherwise it will swell, crack and fall off.

This 'conventional' slaking is actually another form of craft-based hydration to powder. At this time slaking with sand to form mortar, as

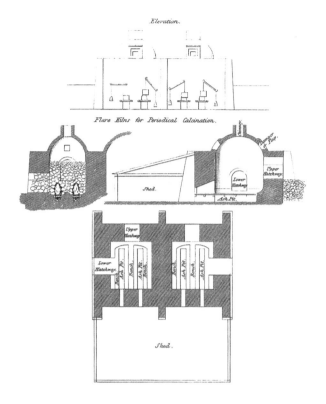

Figure 12 Kilns used for burning blue lias lime at Ramsgate. (Courtesy of the RIBA Library.)

performed by Mr Watson, would in reality have still been by the more conventional method – though use of a pugging mill instead of mixing and beating by hand would no doubt have saved much time and effort, and helped to ensure a much more consistent product.

> Three parts of lime, four of sand, four of coal ash and two of pozzolana will make a good bluish-lead coloured mortar.[35]

The latter mix would in all likelihood have been an effective hydraulic mortar for this time.

The following mixes are normally used:

Brickwork below water: 1 part lime to 1 or 1½ sand

Brickwork above water: 1 part lime to 2 parts sand

Plaster base coat: 1 part lime to 3 parts sand

Plaster finish coat: 1 part lime to 2 parts sand

Concrete: 1 part lime to 6 parts sand is ample.[36]

Donaldson notes that that blue lias lime in these mixes was widely used by Smirke at the British Museum, including the internal plasterwork

Figure 13   Entrance hall, British Museum: view of the main staircase by L. W. Collmann, 1847. Collmann and Davis were responsible for carrying out the decorations to the Museum during the last phase of its construction, the southern entrance range built from 1841 to 1848. (Courtesy the Trustees of the British Museum.)

(Figure 13) where there were initial problems because the material behaved very differently to Medway or Dorking lime, and practice was therefore required by the workmen. Unfortunately he was not specific about what form the differences took.

Between its relatively late introduction (partly a result of Smeaton's delay in publishing his findings), the invention of Roman cement and the overall success of high-fired Portland cement from around 1850, the heyday of blue lias lime within London was very short and its true potential was never fully realized. It is a great pity that the recent re-introduction of the material from Charlton Adam in Somerset was so short-lived, as its potential for use in the conservation of late eighteenth and early nineteenth-century buildings barely had sufficient time to be recognized.

Cements for engineering structures

Donaldson notes that artificial cements were available in 1840 but that they were still considered inferior to natural cements; presumably this

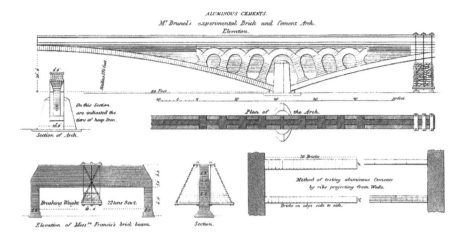

Figure 14   Illustration from Donaldson of a range of structural tests carried out in the years immediately prior to 1840. (Courtesy of the RIBA Library.)

situation continued until the introduction of high-fired Portland cement in 1845–50 by William Aspdin. Donaldson includes an illustration (Figure 14) in his article of several structural tests recently carried out using available good quality natural cements, although he does not discuss these in great detail. The performance of several of these materials was sufficiently dramatic to have been at the time the subject of a great deal of enthusiastic discussion.

The illustration at the bottom right in Figure 14 shows an early method of testing the adhesive and tensile properties of cements. Although this method is not discussed by Donaldson, a similar process is described in detail by Pasley (though not illustrated). The diagram shows 26 bricks projecting horizontally out from a wall held simply by the cement, illustrating behaviour that no lime could possibly match. Pasley notes that when tests were carried out with slower setting 'Metallic Cement' by Mr Troughton, the bricks might be held in place with a board until the mortar had cured sufficiently, after which the projecting horizontal 'pier' was loaded up with bricks on its end:[37]

> Two bricks were fixed to a wall, by joints of Sheppy or Harwich cement, to which 7 others were attached by the mixtures of blue lias or of Merstham lime with copper slag, and all supported by a strong plank as aforesaid, until the time of trial, after which each of these little piers projecting horizontally was loaded with bricks piled up vertically upon the extreme end of it, until it broke down. When thus loaded, the brickwork projecting from the wall had the appearance of the capital letter L.

Figure 14 (bottom left) illustrates a 'brick beam' produced by Alfred Francis's company to demonstrate what their product ('Medina Cement') could achieve. It appears that the success of this demonstration inspired Brunel to use this material for the test shown in the upper illustration. This was a trial construction carried out to validate the principles he intended to use in building his bridge across the Thames at Maidenhead for the Great Western Railway in 1838. The proposed spans of this bridge were by far the widest and shallowest ever seen and well outside the long-established proportions for functioning arches. It is rumoured that the railway

Figure 15   Brunel's bridge over the Thames at Maidenhead for the Great Western Railway. The bridge has been widened since originally constructed to take four instead of the original two tracks, but its elegant profile remains. The original cement mortar can be seen under the bridge, and remains in surprisingly good condition.

company refused to accept the bridge at first, demanding that the timber formwork should be left in place after its construction.

As can be seen in Figure 15, the Maidenhead bridge was never intended to perform as an arched structure in the conventional sense and the test shows a structural principle based on a balanced pair of cantilevers. According to Charles Pasley, who discusses this structure more fully, the brickwork was held together by neat Francis's cement, and hoop iron was laid in the upper joints to give tensile strength.[38]

This was a pioneering type of structure, achievable only in cement. Although arched in shape and constructed of brick, it was designed and built to perform in a manner that was unprecedented in any of the many thousands of true arches using lime-based mortars that had been constructed from ancient Roman times on. This was a significant moment, which marks the beginning of modern practice, achieving the same for structural engineering that Roman cement achieved for architecture and general building.

It is not surprising that, as a result of extraordinary performances like these, lime began its retreat from the mainstream building industry in the face of the overall rise of cement. This process truly began around 1850 and was completed perhaps a hundred years later, in the aftermath of the Second World War, with the final establishment of the now accepted conventional practices of the modern building industry. While this relegated the softness, flexibility and breathability, as well as the aesthetic qualities of lime primarily to historic buildings, it did enable the remarkable engineering and architectural achievements of the later nineteenth and twentieth centuries, including the Victorian and Modern eras.

EPILOGUE: A CONTINUING TRADITION

The use of cement for civil and structural engineering projects proceeded rapidly from these initial demonstrations of its amazing abilities, though it was not adopted so quickly for general building.

Figure 16 shows two houses built respectively in 1908 and 1911 on the Brentham Estate in West London. This was the world's first garden suburb, consisting principally of a range of simple, unpretentious, art-and-craft cottages. The brickwork is constructed using a grey chalk lime mortar, with small amounts of brick dust and ash added to assist the set and to slightly modify the colour. A certain amount of original Portland cement

Figures 16   (a) No. 18 Ludlow Road and (b) Nos 4/6 Meadvale Road, Brentham, West London.

was used on these houses, however, of a strength that subjectively appears to be similar to a strong modern NHL3.5 or weak NHL5 hydraulic lime, but it is placed only where its properties are required such as for pointing creasing tiles and roof verges, bedding copings and as a basis for the roughcast render (originally all unpainted).

This type of construction was entirely unremarkable at the time. It builds on an established regional tradition going back to the seventeenth century and earlier, but takes advantage of relevant developments over the following centuries and the materials used are selected from a position of knowledge to be the most appropriate and effective for their purposes. In the face of what we now know about the relative characteristics and shortcomings of both old and new materials, such a balance seems entirely appropriate and well considered. It is entirely within the normal scope of architectural conservation that such variation is appreciated and replicated, but the resulting qualities are such that we are now starting to see this type of approach returning (albeit tentatively) to the wider construction industry. The likely improvements in the appearance, performance and durability of new buildings which can result should be warmly welcomed.

This paper was originally published in Volume 16 of the *Journal of the Building Limes Forum*, 2009. This is a revised version.

Notes

1 Smeaton, J., *A Narrative of the Building and a Description of the Construction of the Eddystone Lighthouse*, G. Nicol, London, 1793.
2 Holmes, S., 'To wake a gentle giant – grey chalk limes test the standards', *Journal of the Building Limes Forum*, Vol. 13, 2006, pp. 9–20.
3 Smeaton, *op. cit.*, 1793.
4 Lynch, G., 'The myth in the mix: the 1:3 ratio of lime to sand', *The Building Conservation Directory*, Cathedral Communications, Tisbury, Wiltshire, 2007.
5 Smith, P., *Rivington's Notes on Building Construction*, Volumes I–III, Longmans, London, 1904. Reprint with an introduction by Hurst, L, Donhead Publishing, Shaftesbury, 2004.
6 Donaldson, T. L., *Lime, Mortar, Stucco, and Cement: Being an Article, Headed Stucco, in the Volume of Miscellanies in the Encyclopaedia Metropolitana*, Benjamin Fellowes, London, 1840.
7 Dearn, T. W., *The Bricklayer's Guide to the Mensuration of all Sorts of Brick-Work, ... with Observations on the Causes and Cure of Smoky Chimneys, the Formation of Drains, and the Best Construction of Ovens*, Josiah Taylor, Architectural Library, London, 1809.
8 Vicat, L. J., *Recherches Expérimentales sur les Chaux de Construction, les Bétons et les Mortiers Ordinaires*, Goujon, Paris, 1818.
9 Higgins, B., *Experiments and Observations made with the View of Improving the Art of Composing and Applying Calcareous Cements, and of Preparing Quick-Lime: ... and Specification of the Author's Cheap and Durable Cement for Building, Incrustation, or Stuccoing, and Artificial Stone*, T. Cadell, London, 1780.
10 Ibid.
11 Dearn, *op. cit.*
12 Smeaton, J., *Reports of the late John Smeaton FRS, made on various occasions in the course of his employment as a civil engineer*, 2nd edition, two volumes, M. Taylor, London, 1837. (First published in an abridged, limited edition in three volumes: volume 1 in 1797 and volumes 1, 2 and 3 in 1812.)
13 Vicat, L. J., translated in English by J. T. Smith, *A Practical and Scientific Treatise on Calcareous Mortars and Cements, Artificial and Natural*, John Weale, Architectural Library, London, 1837.
14 Pasley, C. W., *Observations on Limes*, 1838. Reprinted with introduction by M. Wingate, Donhead Publishing, Shaftesbury, 1997.
15 Smeaton, *op. cit.*, 1793.
16 Smeaton, J., *John Smeaton's Diary of his Journey to the Low Countries 1755*, The Newcomen Society, Leamington Spa, 1938.
17 Ibid.
18 Smeaton, *op. cit.*, 1793.
19 Pasley, *op. cit.*
20 Smeaton, *op. cit.*, 1837.
21 After Smeaton, *op. cit.*, 1793; modified Holmes, *op. cit.*
22 Donaldson, *op. cit.*

23 Pasley, *op. cit.*
24. Holmes, *op. cit.*
25 Donaldson, *op. cit.*
26 Ibid.
27 Donaldson, *op. cit.*
28 Pasley, *op. cit.*
29 Ibid.
30 Lynch, *op. cit.*
31 Donaldson, *op. cit.*
32 Pasley, *op. cit.*
33 Donaldson, *op. cit.*
34 Ibid.
35 Donaldson, *op. cit.*
36 Ibid.
37 Pasley, *op. cit.*
38 Ibid.

The History, Use and Analysis of Roman Cements

*David Hughes, Simon Swann, Alan Gardner
and Vincenzo Starinieri*

THE ORIGINS OF ROMAN CEMENT

Roman cement and limes

The eighteenth century witnessed substantial developments in the understanding and provision of cementitious materials, the first since the advances made by the Romans. In the middle of the century Smeaton undertook a search for suitable hydraulic limes to use in the construction of the Eddystone lighthouse, and in the process confirmed the link between binder performance and the proportion of clay in the source limestone.[1]

In 1796, a patent was granted to the Rev James Parker for his invention of 'Roman cement', a material which, unlike limes, required grinding rather than slaking prior to use. It was notable for having a brown colour and rapid setting characteristics; the latter typically taking some fifteen minutes. This soon led to the development of a family of allied cements with varying characteristics, some of which were even more reactive, setting in as little as one minute. It is this European family of 'Roman cements' that is referred to in this chapter.

The pace of technological advance accelerated, and the patenting of Portland cement followed in 1824, with Vicat publishing his famous treatise in 1828. By the middle of the nineteenth century Roman cement was substantially superseded in the UK. While its period of major use was

Figure 1   Kedleston Hall, Derbyshire: the hall was originally constructed in the 1760s and rendered in lime, but this was replaced with Roman cement in the early nineteenth century, lined out and coloured in imitation of ashlar.

relatively short, Roman cement has had a considerable impact on cultural heritage throughout Europe, and interest in its conservation is growing (Figure 1).

Development of stucco

The London Building Acts of 1774 targeted the slipshod construction of party walls, and sought to make the 'exterior of the ordinary house as nearly incombustible as possible'.[2] Together with the development in speculative brick-built, rather than stone, construction, and the high cost of stone and labour,[3] an increased demand was generated for external finishes which would be both fireproof and give the appearance of stone. Architectural practice was also developing, and greater coherence and durability was expected of stuccoes on taller and more exposed façades and parapet walls. Stucco was also expected to satisfy the needs of projecting elements previously executed in stone, such as cornices, string courses and drip details, in addition to a range of applied ornamental features.

In the eighteenth and early nineteenth centuries, common stucco in London and south-east England was dominated by the feebly hydraulic limes of Dorking, Merstham, Godstone and Halling.[4,5] Gwynn, being a

strong advocate of the use of stucco, had in 1766 lamented the poor materials available in London, and identified the need for 'stucco resembling
stone, more durable than the common sort'.[6] Yet in 1835 Nicholson was still
reporting that few stucco formulations exhibited long-term durability.[7]

Roman cement stucco

Donaldson reported in 1847 that the more eminently hydraulic blue lias
limes were under-exploited in London, making a late appearance to this
particular market, and this enabled Roman cement to become established locally.[8] Telford recommended its use as stucco, if mixed with a
'proper lime mortar', in situations where previous problems had occurred
through the effects of dampness of the wall.[9] Its rapidity of setting and low
shrinkage made Roman cement an ideal material for the production of
various cast and in situ elements. These included balustrades and pilasters,
where the cement was applied to a brick or tile substrate to build out the
form of such elements.

Roman cements in new construction types

It was also noticed that the properties of Roman cement might allow
entirely new methods of construction. Two notable 'show houses' were
built in the early nineteenth century to demonstrate the possibilities
offered by the use of cement: Castle House in Woodbridge, Suffolk, built
in 1805–09 by William Lockwood; and Castle House in Bridgwater,
Somerset, built *c.* 1851 by John Board (for William Ackerman), and
presumably also intended to show off his products (see below).

The Woodbridge example is assumed to have been the more conventional of the two, being mainly applied render, but with a highly decorated
grotto in the grounds which exhibited ornamental, and probably cast, work
by James Pulham, who worked for Lockwood at the time. It was, however,
noted for its use of cement as a roof covering, although this apparently failed
quite quickly due to the movement of the wooden rafters beneath.

The Bridgwater house still exists, and exhibits a range of notably
innovative constructional features, such as an extensively pre-cast Roman
cement façade, including rusticated blocks and most of the window
elements (Figure 2). The flooring system also appears to be very advanced.
Significantly, the Roman cement used by Board was based on blends of
stones from specific beds in the quarry at Puriton, near Bridgwater, and
this allowed the cements to yield the colours of Portland or Bath stone.

Figure 2   Castle House, Bridgwater, Somerset: detail showing the principal construction elements on the façade. All are in cast Roman cement, and the rusticated wall units are hollow. The lighter colour of Board's cement can be clearly seen by comparison with Figure 4.

WHAT IS ROMAN CEMENT?

The materials

Limes and cements can be broadly classified into the following main categories:

- Limes – derived from limestones that are fired in a kiln and slaked with water to produce limes. These can be further categorized as pure limes or hydraulic limes.
- Calcareous cements – derived from marls or blended limestone and clay mixtures (or other calcareous and argillaceous materials), which when calcined require grinding rather than slaking before use; due to their relatively low free lime content they will not slake to a powder. These can further be usefully divided into natural cements, early artificial cements (using low temperature calcination) and Portland cements (using high temperature calcination).

Roman cement is the frequently used generic term to cover a range of natural cements produced from marls or septaria: these are limestones containing 25% or more clay. Roman cement is classified as 'natural', as all the necessary oxides (lime, silica, alumina etc.) are intimately mixed in the one source material, rather than being blended from different sources, as is the case in the production of Portland cements. Each Roman cement will therefore reflect the local geology from which it is sourced, yielding a range of setting times, setting speeds and colours across the genre.

While both Roman cement and natural hydraulic limes are calcined at relatively low temperatures, they are differentiated by the rapid setting exhibited by the cement.[10] Both materials contain substantial quantities of belite (dicalcium silicate; C_2S). However, Roman cements develop a significant early strength in 1–6 hours, but can then undergo a dormant period of up to 1 month before the belite hydrates to give the expected prolonged strength development profile. Roman cements can be differentiated from Portland cements by the presence of residual quartz and calcite, and the absence of alite (tricalcium silicate; C_3S), which is responsible for the substantial early age strength of Portland cements.

Work undertaken during the ROCEM project on newly produced Roman cements has indicated that its rapid setting is a function of the production of calcium aluminate hydrates, but the mineralogical source of the aluminate is uncertain.[11] The only commonly identified crystalline aluminate is gehlenite, but this is normally considered unreactive. Analysis of the mineralogical and oxide contents reveals a substantial but unidentifiable phase which is amorphous to X-ray analysis. The principal oxide components of this phase are lime, silica and alumina, and it is now believed that such 'amorphous' or micro-crystalline alumina is responsible for the rapid setting of Roman cements.

History of Roman cement production

A description of the discovery of 'Roman cement' by the Rev Dr James Parker can be found in a pamphlet published in 1832, attributed to an anonymous civil engineer.[12] The derivation of the name is a classic case of the inventor wishing an association to be drawn between their new material and those of proven quality – specifically, the ancient Roman lime and pozzolan blends. Several authors, such as Pasley, were highly critical of the use of the name, and recommended referring to each cement by its source.[13] However, the practice of blending cements was not uncommon, and Pasley himself claimed that a cement blend produced from a combination of Harwich and Sheppey cement stones (referred to as English cement by its manufacturer, Francis and Sons) was better than either of the purer products.[14,15]

Small-scale kilns were common in Roman cement production in the UK, and frequently used coal, coke or timber as their fuel. The only pre-processing undertaken was to reduce the stones to fragments the size of a fist, a task often left to young boys.[16,17] It is clear that little attempt was made to remove the extraneous clay adhering to many of the stones, and the use of a riddled shovel was recommended to separate calcined stone and clay when the kiln was being unloaded.[18] European production of Roman cement commenced several decades later than in the UK, and industrial scale kilns were therefore more common from the outset.

After some initial confusion it is now clear that the necessary calcination temperature is low, being similar to that required for limes, and there is no need for temperatures 'nearly sufficient to vitrify' as was claimed by Parker. The comments in contemporary literature that high calcination temperatures yielded inferior cements have been confirmed, and it is now recognized that most kilns of the day would yield a range of temperature profiles within a single batch of cement.[19]

It would appear that the initial take-up of Parker's cement was slow, and he sold his patent in 1798 to a partnership of Samuel and Charles Wyatt. However, by the time of the expiry of the patent, others had identified alternative sources of suitable materials, and sites manufacturing Roman cement soon flourished, particularly around the English coast stretching from the Isle of Wight to Whitby.

The first major competition arose following James Frost's discovery of suitable raw material on the Essex coast near Harwich. Here, the material was available both as septaria and in three to four solid layers below beach level. These materials yielded a dark chestnut brown cement. By the 1830s, the industry was using some 30–40,000 tons of Harwich stone annually, which was obtainable at a quarter of the price of septaria from Sheppey.[20]

The co-burning of stones from Harwich and Swalecliffe yielded a similar colour to the Sheppey cement. The addition of Swalecliffe stones also produced a slower setting cement by moderating the exceedingly rapid set of the pure Harwich product, although its final strength was not as great as that of the cement produced from Sheppey septaria alone.[21,22]

Other sources, such as the area around Whitby and Speeton on the Yorkshire coast, produced cements of lighter colours, more readily akin to those of the prized Bath or Portland stones, and were as a result greatly favoured for stucco. Production of Whitby cement was commenced at Eastrow, Sandsend in 1811, and was variously known as Mulgrave, Yorkshire, Whitby or

Figure 3   The kiln at Sandsend near Whitby used to produce Mulgrave cement. The cement was stored in barrels made in the cooper's shed at the rear of the kiln.

Atkinson's cement, although this production was only viable when associated with the local alum industry (Figure 3). Further down the coast at Hull, George and Thomas Earle calcined stones from both the Mulgrave Estate and Speeton, commencing in 1821. Subsequently, they also used Harwich stones and sold both 'Light' and 'Dark Cement'.

Medina cement was produced on the Isle of Wight from 1840, originally using stones sourced from the Hampshire coast near Christchurch, before changing to stone dredged from the sea off Kimmeridge. Millar reports Medina cement setting 'almost as soon as it leaves the trowel', so making it ideal for cast work.[23] It would appear that the Medina and Mulgrave cements were of a similar colour.

John Board produced various cements from 1844 at his Dunball works near Bridgwater in Somerset. Not only was it claimed that these cements resembled Bath and Portland stones, but also that they would take more sand than any other cement.

Composition of Roman cements

It is clear that each historic source yielded Roman cements with particular properties. Contemporary analyses reflected both this variation, and the inherent variation within each source. These analyses, however, are limited compared to modern techniques, as they were restricted to considering either oxide compositions or limestone and clay proportions. More significant information is available to us via the use of X-ray diffraction (XRD) techniques (Table 1), which show the actual clay type and, in combination with an oxide analysis, can reveal how much of the silica is present within the clay and quartz separately.

Table 1   Mineralogical analysis of septaria from various sources.

	Quartz	Feldspar	Calcite	Pyrites	Clay minerals
Sheppey	19	1	61	2	(16% illite + 1% kaolinite)
Harwich	9	2	61	2	(16% smectite + 7% illite + 3% kaolinite)
Whitby	10	0	64	2	(7% illite + 18% kaolinite)

HISTORIC APPLICATIONS AND MIXES

Principal applications

The principal uses of Roman cement were for stuccoes, cast decorative and functional elements, and specialist building mortars. In addition to external stuccoes, it was used for internal hardwearing plaster, particularly in cellars or areas of buildings to which the public had ready access.[24] It was occasionally used in concrete – Medina cement being used in house construction, whilst it has also been reported in the foundations of a road on the Highgate Archway Trust.[25] As a consequence of its rapid achievement of coherence, it found use in the construction and repair of hydraulic structures as well as for fixing stonework and terracotta components.

Methodology of application and construction of render

The development of Roman cement allowed a range of new techniques and methods of application that were not possible with earlier materials. We are consequently still developing an increasing understanding about many of these with each investigation made (Figures 4 and 5). The critical

Figure 4   Hadlow Tower, near Tonbridge, Kent: constructed from 1838 using Roman cement in a range of techniques as an integral part of the design.

Figure 5   Hadlow Tower, near Tonbridge, Kent: cross-section through a cast ornament showing a fine-grained surface coating on a coarser base layer, bulked out with tile pieces.

aspects were often the fast set of Roman cement; its nearly instant initial strength gain; its tenacious bond or adhesion, and the ability to work fresh on fresh with successive coats without significant shrinkage. The range of possible new techniques included the following:

- the building up of 'backings' or cores for projections, pilasters or similar forms, typically with tiles bonded to the wall with cement;
- in-situ run and bench run applied mouldings;
- extensive castings, often with two layers, a finer outer layer and a coarser core, sometimes with brick or tile integral to the casting;
- concrete backings and infillings of areas to be stuccoed (found only occasionally);
- projecting and cantilevered brickwork often forming a core for stucco application. Such cantilevers often rely solely or extensively on the tenacious adhesion of Roman cement to support the construction;
- flat renders;
- pre-cast concrete blocks as seen at Castle House, Bridgwater, and in widespread use in the architecture of Grenoble.

Ferrous metals

Iron was used extensively in many Roman cement stuccoes, particularly in the form of nail armatures for run mouldings, sometimes with twine (string) tied between the nails for additional support ('spike and rope

Figure 6   Pinnacles at Hainford Church, Norfolk: the principal mouldings on the pinnacles were either pre-cast or bench run and applied. The rest of the building was constructed with lime mortars.

Figure 7   Norwich Castle, Norfolk: Roman cement render lined out in imitation of ashlar and in-situ run hood moulds.

Figure 8   Nowton Church, Bury St Edmunds, Suffolk: Roman cement cast elements for reredos.

Figure 9   Norwich Castle; 'spike and rope' bracket armature, showing results of expansive corrosion of the iron spike.
Figure 10   Newnham Terrace, Cambridge: damaged console bracket exposing ferrous armature.

brackets' as described by Millar).[26] This type of armature has been found at Wycombe Abbey and in the Prison Blocks at Norwich Castle (Figures 7 and 9). However, hoop iron and round bar can also be found, both in cornices and more typically as a method of fixing cast elements.

Most ironwork is now either causing cracking and delamination through expansive corrosion, or is almost certain to cause it in the near future (Figure 10). Iron and steel within or behind Roman cement should therefore, wherever possible, be removed and replaced with stainless steel. If this is impossible without causing excessive loss to valuable historic fabric, then other forms of protection from rusting must be considered.

Historic mix specifications

A range of historic mix specifications may be found in documents of the period, generally depending upon the particular application.

Mortars for general building work:

> For general building work, proportions of 1 : 1 of Roman cement to washed sharp sand were typically recommended, although the cement content could increase if particularly stringent conditions were encountered.[27,28,29]

Pasley recorded Brunel's specifications for the Thames Tunnel to be 1:1 in the foundations and lower parts of the Tunnel; 1:½ in the piers which support the arches, and neat cement above the springing.[30] Burn provides examples of the leanest mortars of 1:1½, and even 1:2 being used.[31] Donaldson also recommended a maximum bed joint thickness of half an inch (12.7 mm).[32] On occasions, a blend of Roman and Portland cements was specified, such as at the Grosvenor Hotel, where the setting of Portland cement was accelerated by mixing one part of Sheppey cement to three of the Portland.[33]

Stucco mixes:

For stucco use, mixes varying between 1:¼ and 1:1½ are recommended in the contemporary literature.[34,35,36,37,38,39] While it was common to recommend that stucco be applied to a wetted substrate (e.g. Burn), the contrary view may also be found.[40,41] Burn additionally recommends that the surface of the stucco be worked only to yield the 'grain of sandstone'.

Civil engineering mixes:

There is general agreement in the literature that Roman cement was best used neat for hydraulic engineering works.

Mixes for decorative elements:

It is perhaps strange that there is a scarcity of specifications for cast decorations, although this may have been left to the discretion of the individual producer.

In the preface to his book, Burnell argues that while professional engineers and architects may specify mix proportions, the final arbiter is the 'foreman of the pug-mill', who mixes to 'suit his fancy'.[42] While this is likely to be at least partially true, the earlier *Dictionary of Architecture* suggested that the final composition would depend on both the cement and the sand type, and this is under current investigation within the ROCARE project.[43]

Sand grading:

Although sand was commonly specified to be 'good, clean and sharp', it is very rare to find any reference to its grading. Notable exceptions, although still vague, include some specifications cited by Donaldson which stipulated

Thames sand 'from above the bridge' intended for use in mortars for structures in the metropolis.[44] Burn reports Vicat and Treussart recommending
fine sand for use with hydraulic lime and coarse sand for non-hydraulic
lime.[45] Whilet there is no specific evidence of such general guidance being
transferred for use with Roman cement, craftsmen of the time were familiar
with a wide range of materials. They may therefore have made connections
for themselves and selected fine sands for use with Roman cement, which
was sometimes thought of as simply an extension to the range of eminently
hydraulic limes.

Adhesion between layers

The historic literature does not recommend the use of more than a single
layer of Roman cement mortar; there are, however, many examples where
this has been used. In the experience of one of the authors, this accounts
for the majority of renders examined, although it should be recognized
that these are likely to have been 'fresh-on-fresh' application rather than
incremental layering on well-cured earlier layers.

Examples include the pinnacles of Hainford Parish Church, Norfolk
(1838–40 by John Brown) (see Figure 6) and the New Prison at Tothill Fields,
Westminster (1830 by Robert Abraham). In the former case the base layer has
been constructed using the dark brown Harwich cement with the outer layer
being a lighter brown, probably achieved by the addition of lime rather than
the use of the lighter Sheppey cement. In contrast, the coping to the Dart-
moor granite wall of the prison was bedded in Roman cement and pointed
in Atkinson's cement.[46] Presumably the issue was colour acceptability in
both cases and subsequent layers added after the first had set.

Craft practices could be adopted to enhance the potential for layer
adhesion. These may include the application of slurry or splatter coats,
scratch keys and other forms of mechanical keying.

Historic surface finishes

Many Roman cement renders would have been given an applied finish,
often to disguise their colour if this was considered inappropriate. Inevitably, these have often become obscured or eroded over time, but they are
nevertheless clearly a vital part of the value of the building, and need to
be understood. It was also common to work finishes into the surface, such
as struck jointing in imitation of ashlar, which will often integrate with
moulded and cast detail.

Various techniques for tinting or colouring the surface of Roman cement renders were developed, particularly to help the material more closely represent stonework. The simplest approach was to use a limewash or lime–cement wash that was repeated over the whole façade in order to present a uniform appearance. Quicklime was a common base, supplemented with copperas to modify the pure white of the lime, yielding a rich yellow-gold Bath stone colour. Other potential ingredients included beer grounds (which Pasley did not recommend), milk, alum, tallow, lamp black, spruce yellow ochre and even cement (Figure 11).[47]

For optimum colour retention limewashing was best performed within one hour of rendering in order to form a bond with the cement.[48,49] This procedure had the additional advantage of covering the fine shrinkage cracks common in Roman cement façades.[50,51]

Nicholson describes a technique known as fresco in which a dilute solution of pigmented sulphuric acid was washed onto the surface of the limewash to reduce the uniformity of appearance, and yield the 'in imitation of stone' requirement which was often specified.[52,53] It was sometimes recommended that the ashlar struck joints should be additionally tinted to mimic a traditional masonry finish.

The disadvantage of limewash and fresco-based systems was the need to re-apply the wash to combat the effect of the weather as frequently

Figure 11   A Roman cement casting disfigured by a thick coating of a cement spray.

as every 2–3 years, although Bristow suggests that this may have been extended to a 4-year cycle to coincide with the normal frequency needed for repainting of wood and ironwork.[54,55,56]

Oil-based paints were also commonly used on Roman cement renders, particularly in large towns where they helped to resist soot and pollution. These gave a less natural-looking finish, but had the advantage of extending the redecoration cycle to perhaps 5–7 years. Such finishes are almost a hallmark of early nineteenth-century high-status buildings, such as Nash's terraces in Regent's Park in London.

This process, however, will have often led to a considerable build-up of paint over time, sometimes aggravated by modern masonry paints as the top layer, which will significantly obscure the original decorative detail. In some instances crude Portland cement slurries or mortars may have been applied as new surface finishes.

ANALYSIS OF NINETEENTH-CENTURY MORTARS

Mortar mixes as found in current analyses

The composition of mortar samples from some 50 structures in the UK, Poland, Austria, Germany, Slovakia and the Czech Republic have been analysed as part of the ROCEM project.[57] A wide range of aggregate content was identified, reflecting significant variations in local practice and preference.

Nevertheless, run and cast elements typically possessed aggregate contents of 20–25% by weight of the total mortar (approximating to a 1 : ¼ mix), while renders and joint mortar revealed twice that concentration (approximately 1 : ¾). Although the sands exhibited a wide range of gradings, they were significantly finer than that of sand currently recommended for use in lime mortars. It is apparent that the grading of cast elements is often bi-modal, with the use of coarse, rounded particles within the bulk of the elements and much finer sand in the surfaces.

Microscopic analyses

As noted earlier, as part of the ROCEM project, microscopic examination of both laboratory calcined marls and the sampled historic mortars has been undertaken. The comparison of these materials gives a better opportunity to describe and define the observations of the historic mortars.

Typical mortars exhibit a significant proportion of nodules which have not reacted despite the prolonged atmospheric exposure; these may be termed under-burnt, well-burnt and over-burnt relicts (Figures 12 and 14). Figure 13 also shows the consequences of poor grinding, resulting

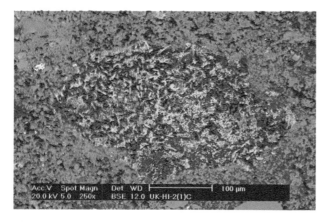

Figure 12 White Lion Hotel, Eye, Suffolk: nodule of under-burnt marl-like remnant.

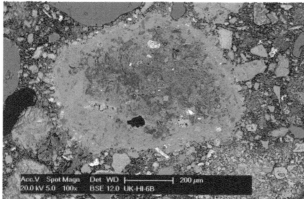

Figure 13 Orwell Terrace Hotel, Harwich, Essex: nodule of correctly burnt but inadequately ground cement, showing reactive phases with only a compact rim of hydration.

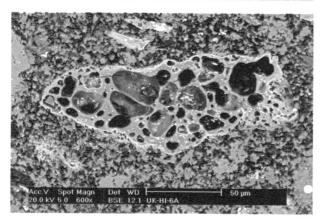

Figure 14 Orwell Terrace Hotel, Harwich, Essex: nodule of over-burnt cement vitrified to glass with voids.

in the hydration of only the rim of an otherwise reactive particle. The largest nodules observed were of millimetre size, which reveals that despite existing specifications for grading, coarse particles were not uncommon. Indeed, Pasley reports that Harwich cement was frequently ground to insufficient fineness, which may well have been the cause of adverse comments on its quality.[58] A full description and characterization of these nodules may be found elsewhere.[59]

Variations in reactive components and grading

It was previously stated that the use of XRD analysis to characterize the marl is superior to a simple oxide analysis (as was carried out historically). This is, in particular, because laboratory calcinations have shown that silica present in the form of quartz does not fully react with lime to form reactive calcium silicates. This is in contrast to silica from clay minerals The presence therefore of a greater amount of quartz is likely to yield a cement with a greater proportion of non-reactive nodules, and these relicts should therefore be viewed as aggregate particles within the cement.

The different Roman cements would also have exhibited different degrees of homogeneity of the original source material, which would inevitably lead to variations in the uniformity of the resulting compositions and particle sizes. Thus, Roman cements as a class clearly often contained their own aggregate, the effect of which will have been exacerbated by poor control of grinding.

The contemporary literature is littered with references to one cement being able to take more sand than its competitors. If sand content is adjusted to suit various cements without resulting in a loss of comparative strength, then, as implied by Spalding, cements with larger proportions of inert materials would have been among those recommended for low sand contents.[60]

Materials analysis for repair

The identification of Roman cements on historic buildings simply by the presence of a dark brown material is unreliable, as Roman cements were produced in a range of different colours, and other materials, even Portland cements, can also be brown.

The most suitable analytical technique for identifying Roman cement is thin-section petrography, supported as necessary by electron microscopy to establish the morphological features and chemical composition of unhydrated cement grains.[61] Data obtained in the latter analysis can then be used to calculate the hydraulic index and cementation index of the unhydrated cement. Thin section analysis is also useful in determining aggregate content and type, the composition of render layers, and whether any micro-structural decay is present within the mortar matrix.

A review of various analytical techniques has recently been published by Varas et al.[62] and applied to Spanish natural cements of the nineteenth century.

REPAIR AND CONSERVATION OF ROMAN CEMENT STUCCOES

Durability

Despite contemporary reports of some early failures,[63] possibly as a result of the use of unsound cement, Roman cement stuccoes have been shown to be very durable. Analysis of historic mortars reveals that they exhibit both high compressive strength and high water transport characteristics.[64] This is an interesting combination, and may well be related to both the pore structure and the morphology of the hydrate structure.[65]

Micro-structural decay

Decay or micro-structural alteration can usually be classified into one of four main categories:

Surface sulfation:

Although Roman cements are generally resistant to sulfate attack, high levels of sulfate at the surface can often be detected in sheltered situations where rain washing has not been effective. The resulting sulfate encrustation and deposition of black surface pollutants can lead to blistering and loss of the surface, especially in the case of cast objects with only a thin original fine finish.

Surface erosion:

Where the stucco has been subjected to heavier than normal weathering, significant surface erosion may have occurred. This can remove fine surface detailing such as imitation 'ashlar' jointing. Coarse aggregate from the core

Figure 15   Hadlow Tower, near Tonbridge, Kent: detail showing damage to exposed surfaces and surface discolouration in more protected areas; note the good state of the in-situ run elements.

Figure 16   Inappropriate use of Portland cement for the repair of modelled Roman cement stucco.

can be exposed to the surface of castings, contrasting with the original finer surface retained on other in situ applied elements (Figure 15).

Aggregate decays:

Significant aggregate decay is rare, but some alteration or minor decay to aggregates is not uncommon over time. Occasionally, where inappropriate aggregates were originally used, decay can lead to expansion and surface cracking (Figure 16). This is visible at Castle House, Bridgwater, where heavily exposed elements have been subject to expansive reactions due to a high sulfate content (fibrous gypsum) in the large laminated lias aggregates used in the casting process.[66]

Carbonation:

Over time, carbonation of cements reduces the alkalinity which protects metals from corrosion. Most historic Roman cement stuccoes are now fully carbonated, and any ferrous metal present within the stucco is therefore liable to expansive corrosion. Carbonation may also affect the stability of some cementitious compounds.

Choice of repair materials

In comparison to the range of hydraulic limes currently available to the conservator, the market has been poorly served in respect of Roman cements. This deficiency was a major driver for both the ROCEM and ROCARE Projects. There have been many examples found of Roman cement inappropriately repaired with modern Portland cements, a technique which is both unsightly and potentially damaging.

Until recently, European conservators had access to only one Roman cement, produced near Grenoble in France. Significant effort has therefore been put into re-learning the production technology, first at laboratory scale but subsequently in pilot plants.[67,68,69] Currently three further suppliers in Poland and Austria are now also producing useful materials in small quantities. While all nineteenth-century Roman cements were produced in vertical shaft kilns, two of these sources have successfully calcined cement in rotary kilns, so refuting the claim that shaft kilns alone are able to produce Roman cements.

Concerns have been expressed that currently available Roman cements are too strong for conservation purposes. Assessment of historic Roman cement mortar samples up to 150 years old, however, has revealed average compressive strengths of 38 MPa (renders) and 44 MPa (castings), although these values were obtained from samples some 15–20 mm thick, so are not directly comparable to the strengths commonly cited from standard 40 mm samples. Nevertheless, it is good conservation practice to use sacrificial mortars in certain circumstances. French Roman cement has been used in blends with natural hydraulic limes. The design of the blend is influenced either by a desire to moderate the strength of the cement, or to accelerate the setting of the NHL, and will depend upon the function of the layer of render under consideration.[70]

Choice of retardation technique

All currently available Roman cements exhibit rapid setting times, generally much shorter than the commonly cited historic standard of fifteen minutes, and the use of retarders is therefore essential.

The most usually adopted technique is to include a small quantity of a chemical retarder, although care must be taken to select one that does not adversely affect the development of strength during curing. Suitable retarders include citric acid, sodium citrate and potassium citrate, although even these in excessive dosages can yield low early age strengths and the cement manufacturer's advice should always be sought.

An alternative approach is to 'pre-hydrate' the mortar under very controlled conditions. While we have not found any references in the UK literature to this technique there is evidence of its practice in Europe. Two reports describe retarding the set of French Vassy cement by mixing it with damp sand and waiting for a short time before using the mortar.[71,72] This advice is at odds with the commonly encountered assertion that only dry sands should be used with Roman cement.

Extensive trials have revealed that a two-stage production process, commonly termed 'de-activation' produces a retarded mortar without compromising the strength performance. By controlling the sand type, de-activation water, de-activation period and mixing time the workable life of mortars can be predicted. Interestingly, this technique also allows the fresh mortar to be rejuvenated at the end of its workable life by simply remixing without the addition of any additional water; a process which does not impact on strength performance.

Techniques for undertaking a de-activation procedure as part of the cement manufacturing process are currently being considered in an attempt to market de-activated cement direct to users. Currently, however, this is recommended only as a procedure for conservation practices with a high skill base.

CONCLUSIONS

Roman cements occupy a position on the hydraulic continuum between natural hydraulic limes and Portland cements, and have a distinctive set of properties. They are essential to the repair of buildings from the first half of the nineteenth century, and their conservation warrants as much attention to detail as would be given to any historic lime mortar.

A fundamental study of both historic Roman cement mortars and laboratory-produced cements has yielded a better understanding of their specification, usage and potential for their conservation, although research is still continuing: the ROCARE web site (www.rocare.eu) provides a wealth of data on the most recent research on both historic and modern Roman cement mortars including a manual of best practice.

This paper was first published in two parts in Volume 13, Nos 1 & 3 of the *Journal of Architectural Conservation*, March & November 2007. This is an edited and revised version.

Acknowledgements

The first two authors are grateful for the financial support afforded by the ROCEM Project funded by the EU under Framework 5 – 'ROman CEMent to restore built heritage effectively' EVK4-CT-2002-00084. The first and last authors also acknowledge support in the 7th Framework ROCARE Project, 'Roman Cements for Architectural restoration to New High Standards', project No. 226898. Thanks are due to Professor Weber at the University of the Applied Arts in Vienna for the electron micrographs of the historic mortars.

Notes

1   Smeaton, J. A., *A Narrative of the Building and a Description of the Construction of the Eddystone Lighthouse with Stone*, G. Nicol, London, 1793.

2   Summerson, J., *Georgian London*, Pleiades Books, London, 1945.

3   Burn, R. S, *Masonry, Bricklaying and Plastering*, 1871, reprinted by Donhead Publishing Ltd, Shaftesbury, 2001.

4   Nicholson, P., *An Architectural and Engineering Dictionary*, Vol. 1, John Weale, London, 1835.

5   Papworth, W., *The Dictionary of Architecture*, Architectural Publications Society, London, 1892.

6   Gwynn, J., *London and Westminster Improved*, Gregg International Publishers, Farnborough, facsimile of the 1766 edition, 1969.

7   Nicholson, *op. cit.*, 1835.

8   Donaldson, T. L., *Encyclopaedia Metropolitana*, Vol. XXV, London (1845).

9   Telford, T., *A copy of a letter to the Secretary of the British Society*, 1796, p. 16.

10 Hughes, D. C., et al., 'Calcination of marls to produce Roman Cement', ed. Edison, M. P. *Natural Cement STP 1494*, ASTM, West Conshocken, PA, 2008, pp. 84–95.

11 Vyskocilova, R., et al., 'Hydration processes in pastes of several natural cements', ed. Edison, M. P., *Natural Cement STP 1494*, ASTM, West Conshocken, PA, 2008, pp. 96–105.

12 Anon, *Practical Remarks on Cements for the Use of Civil Engineers, Architects, Builders etc.*, second edition, William Gilbert, London, 1832.

13 Pasley, C. W., *Observations, deduced from experiment, upon the natural water cements of England, and the Artificial Cements that may be used as substitutes for them*, Establishment for Field Instruction, Chatham, 1830.

14 Smeaton, A. C., *The Builder's Pocket Manual*, M Taylor, London, 1837.

15 Pasley, C. W., *Observations on Limes*, 1838, reprinted by Donhead Publishing Ltd, Shaftesbury, 1997.

16 Anon, *op. cit.*, 1832.

17 Mitchell, J., *Geological Researches Around London*, Vol. 1, Manuscript of the Geological Society Library, London, pp. 254–8, undated.

18 Anon, *Practical Remarks on Cements for the Use of Civil Engineers, Architects, Builders etc*, second edition, William Gilbert, London, 1832.

19 Hughes, *op. cit.*, 2008.

20 Anon, *op. cit.*, 1832.

21 Anon, *op. cit.*, 1832.

22 Mitchell, *op. cit.*, undated.

23 Millar, W., *Plastering Plain and Decorative*, 1897, reprinted by Donhead Publishing Ltd, Shaftesbury, 2004.

24 Donaldson, T. L., *Handbook of Specifications, Parts 1 & 2*, Atchley, London, 1859.

25 Anon, *op. cit., 1832.*

26 Millar, *op. cit.*, 2004.

27 Donaldson, *op. cit.*, 1845.

28 Pasley, *op. cit.*, 1830.

28 Grizzard, F. J., 'Documentary History of the Construction of the Buildings at the University of Virginia, 1817–1828', PhD thesis, 1996, see Appendix T http://etext.virginia.edu/jefferson/grizzard/appt.html (accessed 19 September 2006).

29 Burnell, G. R., *Rudimentary Treatise on Limes, Cements, Mortars, Concretes, Mastics, Plastering etc.*, 14th edition, Crosby Lockwood and Son, London, 1892.

30 Anon, *op. cit.*, 1832.

31 Burn, *op. cit.*, 2001.

32 Donaldson, *op. cit.*, 1845.

33 Millar, *op. cit.*, 2004.

34 Nicholson, *op. cit.*, 1835.

35 Donaldson, *op. cit.*, 1845.

36 Pasley, *op. cit.*, 1838.

37 Donaldson, *op. cit.*, 1859.

38 Burnell, *op. cit.*, 1892.

39 Nicholson, P., *The New Practical Builder*, Thomas Kelly, London, 1823.

40 Burn, *op. cit.*, 2001.

41 Burnell, *op. cit.*

42 Summerson, *op. cit.*, 1945.

43 *Dictionary of Architecture, Vols 2 and 7*, The Architectural Publication Society, London, 1853–92.

44 Donaldson, *op. cit.*, 1859.

45 Burn, *op. cit.*, 2001.

46 Donaldson, *op. cit*, 1859.

47 Pasley, *op. cit.*, 1830.

48 Anon, *op. cit.*, 1832.

49 Donaldson, *op. cit.*, 1859.

50 Pasley, *op. cit.*, 1838.

51 Donaldson, *op. cit.*, 1859.

52 Nicholson, *op. cit.*, 1835.

53 Donaldson, *op. cit.*, 1859.
54 Pasley, *op. cit.*, 1830.
55 Bristow, I. C., 'Exterior renders designed to imitate stone: a review', *Annual Transaction*, Association for the Studies in the Conservation of Historic Buildings, Vol. 12, 1997, pp. 13–30.
56 Bessy, G. E., 'The maintenance and repair of Regency painted stucco finishes', *RIBA Journal*, Vol. 57, 1950, pp. 143–5.
57 Weber, *op. cit.*, 2008.
58 Pasley, *op. cit.*, 1830.
59 Weber, *op. cit.*, 2007.
60 Spalding, F. P., *Notes on the Testing and Use of Hydraulic Cement*, Andrus, Ithaca, New York, 1893.
61 Weber, *op. cit.*, 2007.
62 Varas, M. J., et al., *Natural Cement as the precursor of Portland Cement: Methodology for its Identification*, Cement and Concrete Research, Vol. 35, 2005, pp. 2055–65.
63 Anon, *op. cit.*, 1832.
64 Weber, *op. cit.*, 2008.
65 Swann, S., 'Castle House, Bridgwater, Somerset, Conservators Report for the North and West facades', *Report to SAVE*, October 2003.
66 Bayer, K., et al., 'Microstructure of historic and modern Roman Cements to understand their specific properties', *13th Euroseminar on Microscopy Applied to Building Materials*, Ljubljana, Slovenia, 14–18 June 2011.
67 Hughes, *op. cit.*, 2008.
68 Hughes, *op. cit.*, 2008.
69 Hughes, D. C., et al., 'Roman cements – belite cements calcined at low temperature', *Cement and Concrete Research*, Vol. 39, 2009, pp. 77–89.
70 Sommain, D., 'Technical Specification: Use of Prompt Natural Cement in Mixes with Natural Hydraulic Limes', *The Louis Vicat Technical Centre – Special Binders Section*, 20 June 2006.
71 Chateau, T., *Technologie du Batiment*, Libraire d'Architecture de B. Bance, Paris, 1863.
72 Prevost, J., *Le Ciment de Vassy: Les Travaux en Ciment,* Societe Anonyme des Ciments de Vassy, Paris, 1906.

Materials Analysis, Testing and Standards

COLOUR PLATE SECTION

See the following pages for plates referred to in the text.

Building Limes Standard BS EN 459:2010

Steve Foster

HISTORY OF BUILDING LIME TESTING

Lime has been used for many hundreds of years, but until recent times the quality of lime for building was measured using only the expertise of the craftsman, rather than by modern analytical instruments.

Building lime was historically produced locally, from whatever source of calcium was available, and the properties of the lime produced were inevitably extremely variable. Chalk and limestone of varying degrees of purity were used, and even sea shells if no other sources were available. Product quality testing was basic, and relied heavily on the experience of the lime burner. Experienced men were (and some still are) able to tell the quality of the lime they made from its colour, weight and smell. Good quality, well burnt quicklime has a distinctive smell, and sounds like china when two pieces are knocked together. It is also considerably lighter in weight than the raw stone. The degree of burn can be seen by breaking open the quicklime lumps to reveal the unburnt core, which is darker in colour. Another indication of the degree of burn is the ferocity of the reaction between quicklime and water.

Over time, building lime production has moved from a widespread, local cottage industry to an industrialized, automated process, now carried out on a very large scale by just a small number of companies. Building lime as we now know it is very different to the limes used for building by our ancestors (even as recently as our parents in some cases). Product quality is now measured in numbers by the results of sophisticated testing, rather than by appearance, smell and sound. Modern day test methods and equipment are designed to produce results which are repeatable by staff in the same

Figure 1   Traditional clamp kiln at Melton Ross in the early twentieth century.

Figure 2   Modern Maerz Kilns at Melton Ross, *c.* 2009.

laboratory, and reproducible by laboratories throughout the world. Test methods are well documented, and standard equipment is used.

Lime is now used for a myriad of industrial, chemical, manufacturing and agricultural applications, not just within the building industry, although millions of tonnes of building lime are still sold throughout Europe. However, the majority of this is now sold as quicklime (calcium oxide; CaO) for the production of autoclaved aerated concrete blocks or

Figure 3   Autoclaved aerated concrete blocks (courtesy of H+H UK Ltd.)

Figure 4   Soil stabilization (courtesy of Beach Stabilisation Ltd.)

for soil stabilization or as hydrated lime (calcium hydroxide; $Ca(OH)_2$) for calcium silicate bricks, for cement-based mortars, or for asphalt.[1] It is now important that building limes are produced consistently in order to give predictable performance in these relatively high-tech applications. The British Lime Association website contains information on the wide uses of limes for both building and other applications.[2]

STANDARDS – A MEASURABLE IMPROVEMENT

In 1940, the British Standards Institute introduced BS 890 for Building Lime, which specified chemical and physical requirements for quicklime, hydrated lime and lime putty.[3] The stronger hydraulic limes in the modern NHL and HL strength classes were not included in this standard, although grey chalk limes ('feebly hydraulic' limes) were mentioned.[4] BS 890 was revised in 1966, 1972 and 1995.[5,6,7]

In 1998 the European Committee for Standardization (CEN) decided to amalgamate the Building Lime and Masonry Cement standards which existed in its member countries to create a new, harmonized standard. In May 1989 CEN began the process of drafting this new standard, with the goal of publishing completed national standards in mid 1992. Inevitably, there was some difficulty in reaching a European compromise and the process took much longer than predicted.

A number of important changes were made during this period, perhaps the most important of which was to separate the building lime and masonry cement standards, and to include hydraulic limes within the building lime standard.

The 1995 revision of BS 890 made reference to the work of CEN and the forthcoming EN 459 standard for building lime, although the British Standard still made no explicit reference to hydraulic limes.[8]

EN 459 Building Lime was finally published in 2001.[9] The standard was split into three parts:

Part 1: Definitions, specifications and conformity criteria

Part 2: Test methods

Part 3: Conformity evaluation

Parts 2 and 3 deal with testing for factory production control, and factory inspection for CE marking. Part 1 is the most important part for the user as it contains descriptions of the types of building lime and adopted classes, and the different qualities of each.

Within EN 459-1:2001 there are four types of lime, which are classified as shown in Table 1.

Calcium lime and dolomitic lime are both categorized as air limes because they slowly harden in air by reacting with atmospheric carbon dioxide. Natural hydraulic lime and hydraulic lime have the property of setting and hardening under water. Atmospheric carbon dioxide contributes to the hardening process. The classes of calcium lime and dolomitic lime are defined by their lime content, expressed as a percentage, and the

Table 1   The four principal classifications of building limes in EN 459:2001.

Type of Building Lime	Type Classifications
Calcium Lime	CL90, CL80, CL70
Dolomitic Lime	DL85, DL80
Natural Hydraulic Lime	NHL2, NHL3.5, NHL5
Hydraulic Lime	HL2, HL3.5, HL5

classes of natural hydraulic lime and hydraulic lime are defined by their minimum compressive strength in MPa.

The standardization process comprises a complex series of technical discussions, followed by lobbying and negotiations at national and European levels. The Building Limes Forum and the British Lime Association are represented at the British Standards committee B516/11 'Building Lime', which is in turn represented at the European Standardization committee TC51/WG11 'Building Lime' and its sub-committees. Standards are reviewed every five years, and the review process can take anywhere from one to several years.

It is important to understand that the Building Lime Standard EN 459 is designed to allow manufacturers to make consistent quality

Figure 5   Part of the testing laboratories at Melton Ross. These facilities are essential both for consistency of production and compliance with the Standards.

products, rather than to reflect how those products will perform in use. For example, the strength of a lime mortar made with either lime putty or hydraulic lime will rely heavily, not just on binder strength, but also on the type and grading of sand used, together with mix ratios and good working practice.

The strength measurements used in the Standard are established using a mortar made to a fixed mix, from a standard sand and a predetermined measured amount of water, and cured in specified, controlled high-humidity conditions. The resulting mix is in no way suitable for site use, but it allows consistent and comparable mortar prisms to be made, and thus enables lime manufacturers to compare strengths between different batches of building lime. Laboratory strength measurements are also made at 28 days, whereas the strength of a mortar on site will depend on the drying rate, and will increase significantly up to and beyond 90 days.

The compressive strength bands for natural hydraulic lime and hydraulic lime are based on 28-day results, and have wide tolerances due to inherent limits in the accuracy of the measurement equipment, and the natural margins for error in the test process.

Table 2   Classifications and compressive strengths.

Type of Building Lime	Type Classification	Compressive strength @28 days (MPa)	Vicat classification
Natural Hydraulic Lime or Hydraulic Lime	NHL2 or HL2	≥ 2 to ≤ 7	Feebly to moderately hydraulic
Natural Hydraulic Lime or Hydraulic Lime	NHL3.5 or HL3.5	≥ 3.5 to ≤ 10	Moderately to eminently hydraulic
Natural Hydraulic Lime or Hydraulic Lime	NHL5 or HL5	≥ 5 to ≤ 15	Eminently hydraulic to cement

As a user, it is important to understand the performance characteristics not only of the type of lime one is using, but also of the brand. Products may have a greater or lesser degree of hydraulic set even within a type classification, leading to very different short-term setting and long-term performance characteristics in use. If there is any doubt as to the suitability of a particular product, the lime supplier should be able to give guidance on the product's performance characteristics.

CHANGES IN THE CURRENT STANDARD – BS EN 459:2010

The 2001 version of the building limes Standard has been completely revised to produce the 2010 version. This revision takes account of developments both in products and in the market in the intervening years.

In order to make the standard easier to use, Part 1 (BS EN 459-1:2010) has been separated into two main sections containing *Air lime* and *Lime with hydraulic properties*. These two families of products are then further divided into sub-families as shown in Table 3. Each family of building lime is fully described within its own section of the standard.

Table 3   Families and sub-families of building limes.

Family of Building Lime	Sub-family of Building Lime	Type designations
Air Lime	Calcium Lime	CL90, CL80, CL70
	Dolomitic Lime	DL90, DL85, DL80
Lime with Hydraulic Properties	Natural Hydraulic Lime	NHL2, NHL3.5, NHL5
	Formulated Lime	FL2, FL3.5, FL5
	Hydraulic Lime	HL2, HL3.5, HL5

Air limes

As before, both types of air lime (Calcium and Dolomitic) are defined by their lime content as a percentage: the higher the number, the greater the lime content. The air limes are further sub-divided according to the form of the product: quicklime (Q), hydrated lime (S), lime putty (S PL) or milk of lime (S ML), i.e.:

- CL90-Q is a quicklime with at least 90% calcium and magnesium oxide.
- CL90-S is a hydrated lime with at least 90% calcium and magnesium oxide (pro rata).
- CL90-S PL is a putty lime with at least 90% calcium and magnesium oxide (pro rata).

Although the air lime categories remain relatively unchanged, the scope of the standard now covers all of the following applications:

- preparation of binder for mortar (for example for masonry, rendering and plastering);
- production of other construction products (for example calcium silicate bricks, autoclaved aerated concrete, concrete, etc.);
- civil engineering applications (for example soil treatment, asphalt mixtures, etc.).

Reactivity and particle size distribution of quicklime are also described.

Lime with hydraulic properties

The lime with hydraulic properties portion of the standard has changed significantly. The 2001 version of the standard contains class descriptions and chemical requirements that had the potential to allow cementitious products to be categorized as hydraulic lime (HL). The current revision aims to address this by more clearly defining the sub-families and by requiring compulsory declaration of contents.

Types of lime with hydraulic properties are defined by their compressive strength in MPa: the higher the number, the stronger and faster the product will set. There are now three sub-families of lime with hydraulic properties:

- Natural hydraulic limes   (NHL)
- Formulated limes   (FL)
- Hydraulic limes   (HL)

Natural Hydraulic Lime

Natural hydraulic lime is a lime with hydraulic properties produced by burning of more or less argillaceous or siliceous limestones (incl. chalk) with reduction to powder by slaking with or without grinding. It has the property of setting and hardening when mixed with water and by reaction with carbon dioxide from the air (carbonation).

The hydraulic properties exclusively result from special chemical composition of the natural raw material. Grinding agents up to 0.1% are allowed. Natural hydraulic lime does not contain any other additions.[10]

Natural Hydraulic Limes (NHL) are the preferred choice for building restoration and conservation. They are quarried for their composition, and burnt at lower temperatures in order to promote the formation of dicalcium silicates (belite) which have a slower, weaker hydraulic set than the tricalcium silicates (alite) found in modern cements.

Formulated Lime

> Formulated lime is a lime with hydraulic properties mainly consisting of air lime (CL) and/or natural hydraulic lime (NHL) with added hydraulic and/or pozzolanic material. It has the property of setting and hardening when mixed with water and by reaction with carbon dioxide from the air (carbonation).[11]

Formulated Limes (FL) are mixtures of air lime and/or natural hydraulic lime which may also contain hydraulic or pozzolanic additives from a specified list. The manufacturer is obliged to notify the customer of any cement inclusion, or of any single additive above 5%, or of total additives above 10%. The NHL-Z grade products from the 2001 standard will generally become formulated limes.

These are products designed to give particular characteristics for specific applications, and are likely to be used mainly in new building rather than conservation.

Hydraulic Lime

> Hydraulic lime is a binder consisting of lime and other materials such as cement, blastfurnace slag, fly ash, limestone filler and other suitable materials. It has the property of setting and hardening under water. Atmospheric carbon dioxide contributes to the hardening process.[12]

Hydraulic Limes (HL) are products which, while they contain lime, may also contain cementitious by-products or additions. It is important to note that manufacturers are not obliged to disclose the composition of these products to the customer.

TEST CRITERIA

Part 2 (BS EN 459-2:2010) of the revised standard describes the test methods and equipment used in factory production control. The 2001 standard contained many references to external test method standards, specifically BS EN 196 Methods of Testing Cement. The 2010 standard has replaced many of these external references with integral test method descriptions, so that all the information is now held within the new Standard.

HOW WILL THE STANDARD AFFECT THE USER?

The 2010 version of parts 1 and 2 of the Standard were approved by CEN on 30 July 2010 and were published in September 2010. Following a six-month transition period, they are now mandatory throughout Europe. Part 3 of the standard (BS EN 459-3 Conformity Evaluation) was approved and published during 2011.

The new Standard is expected to make it significantly easier to identify, specify and purchase the type of lime that is needed for any particular use: most building and civil engineering applications are now covered.

Users and specifiers of lime who have particular compatibility requirements, such as for conservation, and therefore need to know what materials any particular product contains, should in all cases choose a natural hydraulic lime (NHL) or a formulated lime (FL). As the speed of set, and the long-term performance characteristics will vary between manufacturers, even for the same lime type and strength class, reference should be made to the manufacturer's or supplier's technical information and advice service. The National House Building Council in the UK has recently published document *NF12: The use of lime-based mortars in new build*, which is a valuable reference document in the first instance.[13]

Lime manufacturers are obliged by the Standard to ensure that natural hydraulic limes (NHL) remain pure building limes, but they also now have a wide range of possibilities to develop blended products for specific uses, and to produce and market these as formulated limes (FL). In each case, however, CE accreditation audits to ensure conformity will now be more frequent, with an audit every 1–3 years, depending on the volume of lime produced. As a result, users may be confident of the quality and consistency of the products they specify and use.

Table 4   Mandatory accreditation frequencies for different volumes of production.

Factory Production Capacity (tonnes)	Accreditation Frequency
> 10,000	Annually
> 1,000 and < 10,000	Every 2 years
< 1,000	Every 3 years

THE CONSTRUCTION PRODUCTS REGULATION

The Construction Products Regulation (CPR) entered into force in the UK and the wider European Economic Area (EEA) on 24 April 2011. On 1 July 2013, it will repeal and replace the existing EU Construction Products Directive (CPD). From that date, CE marking will become mandatory for the majority of construction products marketed both in the UK and wider European Economic Area.

CE marking applies wherever a construction product is covered by a harmonized European product Standard or a European Technical Approval (ETA) that includes conformity procedures, and is evidence that formal declaration and certification of performance have been achieved. BS EN 459-1:2010 *Building Lime*, is such a Standard, and CE marking will therefore be mandatory for all building limes from that date.

BUILDING LIMES – THE FUTURE

Although the new version of the Standard is a significant step forward, there still remains room for improvement. Work began in early 2011 in preparation for the next revision, with a proposal to include tests for properties such as flexural strength. The development of test methods which better suit the particular qualities of limes as replacements for the current cement-based tests is also under active consideration.

Singleton Birch Ltd introduced an 'NHL1' product some years ago to replicate the weakly hydraulic 'grey limes' of south-east England, but this has not been recognized in the 2010 version of the standard, and therefore does not comply. It will not be legal to market this product within the EU after 1 July 2013.[14] There is currently a proposal to introduce a British (only) Standard for products in this category in the interim period, which can be used as a basis to re-apply for the inclusion of this category on NHL in the next revision of EN 459.

The Building Limes Forum and the British Lime Association will continue to work within the standardization process to ensure that future revisions of the standard reflect the needs of the building lime community on behalf of users, specifiers and manufacturers.

This paper was first published in Volume 16 of the *Journal of the Building Limes Forum*, 2009. This is a revised and updated version.

Acknowledgements

The author would like to thank the members of the Building Limes Forum for their valuable input into the standardization process, members of BSI Committee B 516/11 'Building Lime', members of the British Lime Association Products & Standards Committee, the European Lime Association Standardization Committee, CEN TC51/WG11 'Building Lime' and its sub Committees.

Special thanks go to Stafford Holmes, Paul Livesey and Michael Beare of the Building Limes Forum, Darren Scutt, Jacqueline Bancroft, Philip Sheldon and many others from Singleton Birch, and Dr Martyn Kenny of Tarmac Buxton Lime. Finally a hearty thanks to Mike Farey, who has guided me through the world of hydraulic limes with knowledge, expertise and sage advice.

Notes

1 Britpave Soil Stabilisation Task Group website, www.soilstabilisation.org.uk
2 British Lime Association website, www.britishlime.org
3 British Standards Institution (BSI), *BS 890:1940. Specification for Building Limes,* BSI, London, 1940.
4 Holmes, S., 'To wake a gentle giant – grey chalk limes test the standards', *The Journal of the Building Limes Forum,* Vol. 13, p. 9. Also included in this volume.
5 British Standards Institution (BSI), *BS 890:1966, Specification for Building Limes,* BSI, London, 1966.
6 British Standards Institution (BSI), *BS 890:1972, Specification for Building Limes,* BSI, London, 1972.
7 British Standards Institution (BSI), *BS 890:1995, Specification for Building Limes,* BSI, London, 1995.
8 Ibid.
9 British Standards Institution (BSI), *BS EN 459:2001, Building Lime,* BSI, London, 2001.
10 British Standards Institution (BSI), *EN 459-1:2010, Building Lime Part 1: Definitions, specifications and conformity criteria,* BSI, London, 2010.
11 Ibid.
12 Ibid.
13 Yates, T. and Ferguson, A., NHBC Research Paper NF12: *The use of lime-based mortars in new build,* NHBC Foundation, Milton Keynes, 2008.
14 Holmes, *op. cit.*

Sands and Aggregates for Conservation Mortars

Roz Artis

TRADITIONAL AGGREGATES IN HISTORIC LIME MORTARS

The aggregates traditionally used in mortars were normally sourced as locally as possible, and were often very different to those aggregates which are now commercially available. Typical historic sources would have included sea shore sand, river bed sand and naturally occurring sand and gravel pits. Crushed dressings from stone cutting were also used in fine ashlar mortars.

Sand and gravel deposits are found widely throughout the UK, and many are a result of the erosion and transportation of huge volumes of material during glacial ice movements in the past million years. In addition to such large-scale deposits, there are a range of localized sources, such as raised beaches, river gravels and a small number of beach or dune deposits. The complex geology of the UK has meant that sands and aggregates in historic mortars vary widely across the country, and hence there is a huge range in the colour, texture and technical performance of these materials.

Aggregates in the UK mainly consist of limestone, quartz, feldspar and other rock fragments in varying proportions; finer aggregates tend to contain a higher proportion of quartz, and coarser sands a higher proportion of rock fragments. In addition, aggregates taken from some areas may contain varying proportions of shell, coal, earths and clays depending on the particular environment.

Analyses undertaken at the Scottish Lime Centre Trust of traditional lime mortars from all over the UK and Ireland show that locally available sources were almost always used wherever available, saving on cost, transportation etc. These could be screened to remove unsuitable sized fragments from 'as dug' material. Beach sands were commonly used in coastal regions and often contained proportions of shell and coal (particularly on the east coast). These materials have an effect on the performance of the mortar; the shell potentially 'seeding' carbonation, and the coal possibly having some pozzolanic properties.

Even 2000 years ago, the quality of sand was considered worthy of discussion. Vitruvius, Julius Caesar's engineer, stated:[1]

> in buildings of rubble work it is of the first importance that the sand be fit for mixing with the lime, and unalloyed with earth …

> though pit sand is excellent for mortar, it is unfit for plastering for being such a rich quality, when added to the lime and straw, its great strength does not suffer it to dry without cracks.

The following is from Richard Neve's *The City and Country Purchaser and Builder's Dictionary* from 1726:[2]

Sand

1. Kinds.
2. What it is, everyone knows. Its use (in Architecture) is in the making of mortar.
3. There are 3 sorts of sand, Pit sand, River sand and Sea sand.
4. Pit sand is of all the best, and of all pit sand that which is whitest is (by long experience) found to be the worst.
5. Of all river sand, that which is found in the falls of water is the best, because it is more purged. The sea sand is worst of all.
6. The pit sand, because it is fat and tough, is therefore used in walls and vaults. The river sand is very good for rough casting.
7. All sand is good in its kind, if being squeezed and handled it crackles: and if being put upon a white cloth, it neither stains or makes it foul.
8. That sand is bad, which mingled with water, makes it dirty and muddy, and which has been a long time in the air; because it will retain much earth and rotten humour: and therefore some masons will wash their sand before they use it.

Much of that imprecise technical description is what eventually produces surprisingly durable mortars.

Additionally, builders were historically able to screen sands as they were making mortars by slaking quicklime and sand together, effectively drying out the sand, so it could be punched through an inclined sieve (probably made of animal gut woven to different mesh sizes).

The purpose of aggregates in a mortar

The primary purpose of an aggregate in a mortar is as a bulk filler, which can often comprise around 75% of the volume of the mortar. In addition, the aggregate performs a number of other functions:

- it provides a natural colour and texture to the mortar;
- it contributes to the compressive strength of the mortar;
- certain aggregates may assist with carbonation of the mortar;
- certain aggregates may have pozzolanic properties.

MODERN AGGREGATES

For modern building practice, aggregate used within lime mortars should generally conform to a number of requirements. They should not contain: significant proportions of fine-grained material, salts, organic (plant) material, wood fragments, coal, gypsum or other sulphates and iron sulphides. This however can create conflict with the requirements for matching historic aggregates, as they commonly contain one or more of the above. The desire for mortar matching must be balanced with modern building practice requirements and present aggregate availability.

As of 1 January 2004, new European standards were adopted for particle size ranges/grading as set out in BS EN 13139, 'Aggregates for mortars' (with supporting National Guidance given in PD 6682-3). However, the British Standards Institute has retained BS 1199 and BS 1200, the old specifications for sands for mortar and plaster, in their catalogue and available for purchase, although these have been superseded by the new BS EN 13139, as required by European Standards' rules. This new standard is more suited to Portland cement mortars. The old documents have been retained, however, in response to a request by the Building Limes Forum, in recognition of their usefulness in correctly describing the properties of sands for conservation/restoration work with lime binders.

In building conservation work it must be recognized that new work using matching materials may result in the use of aggregates which do not conform to the BS requirements. Indeed Historic Scotland's own guidance (now lapsed) recommends that mortars for re-pointing should 'accurately match the original work in all respects'.[3]

Choosing appropriate aggregates

Matching new mortars to original materials can be achieved by analysing them to determine their constituents. The results can then be used as the starting point for developing appropriate specifications.

Colour is an extremely important factor to be considered when matching aggregates. The aggregate will provide a lime mortar with its natural colour and texture. As the majority of lime binders are white or off-white in colour they allow the sand colours to be more faithfully reproduced than with a cement mortar, which will normally impose its dull grey-blue colour. The colour of a mortar which results from the colour of the aggregate fines can be usefully assessed against Munsell® Soil Colour Charts, which can considerably assist matching.

The type of aggregate which should be used in a mix depends on the intended function of the mortar. In general, aggregates for use in a lime mortar should be well graded, sharp and clean. A well-graded sand is one which has a good distribution of particle sizes, with most particles lying in the very coarse to medium sand range (between 2 mm and 0.3 mm grain diameter). Aggregates for pointing and ashlar work should be finer than those used for bedding and wall core mortars.

Angular ('sharp') sands are preferable to rounded ('soft') sands, as the grains fit well together, interlocking and forming a sound physical structure. Rounded ('soft') sand particles tend to roll over each other, and do not interlock, which may lead to a poorly bonded mortar with a greatly reduced compressive strength. Soft sands are used in cement-based mortars as they provide a greater degree of workability, and the greater strength of the binder can compensate for the lack of interlocking between grains. In lime mortars, however, superior workability is provided by the lime binder itself, so sharp sands can be readily used, providing better physical performance with good workability. The interlocking behaviour of well-graded, sharp sands will allow a lime mortar to achieve significantly greater compressive strength than its binder classification alone might suggest.

Aggregates in a mortar for building and repointing

Sizes and ratios given are approximate and for general guidance only, but as a rough rule of thumb, the maximum aggregate size should be no less than one third of the joint width. Each aggregate type, depending on its sharpness and grading, will have a different binder requirement – it is

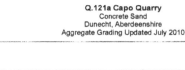

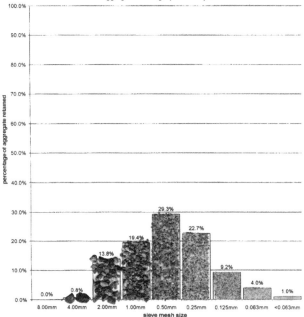

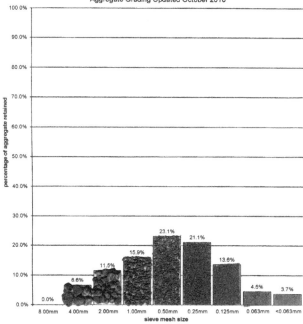

Figures 1 and 2   Examples of well graded aggregates (concrete sands in this case) suitable for use in a lime mortar (sand graphs from Scottish Lime Centre Trust Aggregates Database).

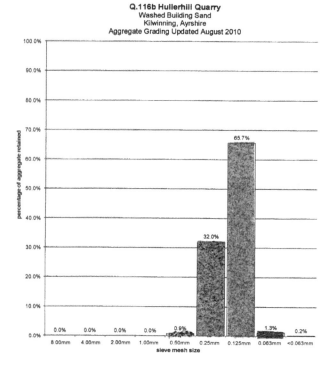

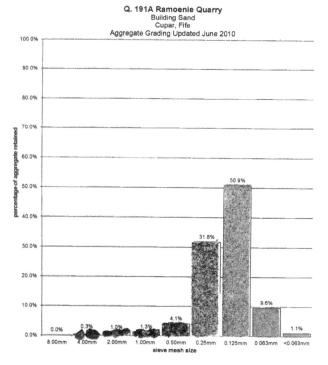

Figures 3 and 4   Examples of poorly graded aggregates, unsuitable for use in a lime mortar. Both aggregates contains a high proportion of 'fines' which will increase the risk of shrinkage cracking in the new mortar (sand graphs from Scottish Lime Centre Trust Aggregates Database).

therefore important that in specifying mortar mixes, this information is known and understood.

Approximate joint widths	Suitable aggregates
Very fine ashlar masonry joints < 3 mm	Crushed stone/crushed chalk
Ashlar masonry joints 3–4 mm	Fine silica/silver sand
Masonry joints 4–8 mm	Sharp building sand
Masonry joints > 8 mm	Sharp concrete sand

Aggregates in a mortar for harling or roughcast

Sharp, well-graded concrete sands are almost always used in harling mortars, except where a flatter finish is specifically required. Grit may also be added to the final coat to achieve a rougher texture although this is a more modern, rather than traditional practice.

As with joint widths for re-pointing, the thickness of harling coats should be no more than three times the maximum aggregate particle size. This will minimize the risk of shrinkage cracking in the new harling. Thinner coats will also carbonate and cure more quickly, and have been shown to be more durable.

Aggregates in a plaster

The scratch and straightening/floating coats will normally use a sharp well graded concrete sand. The sand choice for the finishing coat will depend on the actual finish required.

MORTAR MIX PROPORTIONS

Today we tend to use slaked lime binders (i.e. calcium hydroxide in powder or putty form), and we are likely to use mortar ratios (nominally by volume) of 1:1.5 for 'pricking up' coats, 1:2 for base coats or building mortar and 1:2.5–3 for finish or pointing mixes.

With a basic mortar mix of lime putty and sharp, well-graded concrete sand (coarse stuff), a ratio of around 1 part binder:2.5–3 parts aggregate (by volume) will commonly be required for the binder to fill all the voids between the aggregate particles, and provide a well-bound mortar.

With a basic mortar mix of lime putty and sharp well-graded building sand, a ratio of around 1 part binder:1.5–2.5 parts aggregate (by volume) will commonly be required for the binder to fill all the voids between the aggregate particles, and provide a well bound mortar.

A higher proportion of binder will increase the workability of the mortar, but will also increase the amount of water in the mix, therefore enhancing the risk of shrinkage cracking in the new mortar.

A lower proportion of binder will weaken the mortar and reduce its workability.

In contrast, however, 'hot lime' mortars were often traditionally used, particularly for general building (i.e. quicklime and sand slaked together), and the overall ratio of binder to sand was considerably higher than the figures quoted above. Typical mix proportions in such cases are often closer to 1 part slaked binder to 0.5–1 .5 parts sand (as quicklime roughly increases in volume by 50%).[4]

Slavishly replicating traditional mortars undoubtedly has its dangers, as performance of the original may not have been optimal due to poor quality binders and contaminated aggregates, even without taking into account the environment, location, nature of the repair, intended time of year of working etc, specific to the building. It is therefore normal practice to develop repair mortar specifications which are visually and chemically as close to the original as possible, but modified as necessary to achieve reliable performance.

Analysis of traditional mortars

Traditional lime mortars, in buildings built before the end of the nineteenth century, are commonly composed of a lime binder (commonly non-hydraulic or feebly to moderately hydraulic, to varying degrees) and predominantly silicate or limestone aggregate in varying proportions.

Mortar analysis to determine 'matching' aggregates

Historic Scotland's guidance on re-pointing states that 'any new pointing should accurately match the original work in all respects …'.[5] Sourcing a matching aggregate is commonly the best way to 'match' the colour and texture of a mortar. The ability to match an aggregate accurately will often depend upon the nature of the mortar sample, and the ease with which the aggregate can be isolated from the mortar as a whole for further evaluation.

Basic chemical (acid digestion) analysis

In Scotland, most aggregates are composed predominantly of silicates, which are insoluble in acid. Pozzolanic additives also appear to be uncommon, most likely because the limes tend to be partly hydraulic. This allows a simple visual and chemical (acid dissolution) analysis to be undertaken in most cases.

The process involves dissolving the carbonate binder from the mortar sample with a dilute acid, commonly hydrochloric. This simple chemical analysis can generally be used to derive basic information on the approximate proportions of a lime mortar sample. In its simplest form, it is used to isolate the aggregate from the binder, providing approximate mix proportions. The isolated aggregate can then be further evaluated by grading and microscopic inspection.

In some cases, however, carbonate aggregate may be present which will be dissolved along with the carbonate binder, presenting the main limitation of this technique. Care is therefore required at the outset with acid dissolution, in order to ensure that any carbonate aggregate is identified and quantified prior to acid digestion. Acid soluble aggregates such as limestone and shell occur variably throughout the UK, although the latter can be easily identified by visual examination. In locations such as southern England, where carbonate aggregates and pozzolanic additives are common, alternative techniques of analysis may be more appropriate. [6]

AGGREGATE DATABASES

A database of Scottish Sands and Aggregates is held by the Scottish Lime Centre Trust, and was established in 1998. This database has information on around 600 sands from around 200 commercially operating quarries in Scotland. Work has now started to include sands and aggregates produced in England, Wales and Ireland, thus providing an expanded service to the conservation world. Other reputable institutions which offer a mortar analysis service should also have access to an extensive library of their regionally available sands and aggregates, as without such a resource optimal matching will be difficult if not impossible.

Being able to 'match' sands and aggregates for repair, conservation, maintenance or restoration mortars with lime or early patented cement mortars is important as it will ensure a conservative repair can be effected, whereby only degraded mortars will have to be repaired, limiting the overall volume of work required, and therefore unnecessary intervention in valuable historic fabric.

This paper has been specially written for this volume.

Notes

1 Vitruvius, *De Architectur, c.* 15 BC.
2 Neve, R., *The City and Country Purchaser and Builder's Dictionary*, Sprint, Rivington and others, London, 1736.
3 Historic Scotland, *Memorandum of Guidance on Listed Buildings and Conservation Areas*, Historic Scotland, Edinburgh, 1998.
4 Lynch, G., 'Lime mortars – The Myth in the mix,' *Building Conservation Directory*, Cathedral Conservation, Tisbury, 2007. Reprinted in this volume.
5 Historic Scotland, *op. cit.*
6 Ingham, J., 'Laboratory investigation of lime mortars, plasters and renders', *Journal of the Building Limes Forum*, Vol. 10, 2003. Reprinted in this volume.

Laboratory Investigation of Lime Mortars, Plasters and Renders

Jeremy Ingham

INTRODUCTION

For millennia lime has been used as mortar in the construction of buildings and as finishing materials internally (plaster), and externally (render). The earliest known examples comprise wall and floor plasters from Neolithic sites in the Near East dating from 7000–6000 BC.[1] Construction with lime spread though the Middle East, North Africa, Europe and the Far East, with lime technology apparently developed independently in Central America by the Mayans. The Greeks and Romans refined the use of pozzolana additives to lime, intentionally producing hydraulic lime mortar for underwater works. Following the decline of the Roman Empire the widespread use of pozzolanic sands to create hydraulic mortar largely disappeared.[2]

In the United Kingdom lime:sand mortars, plasters and renders using 'fat' lime (non-hydraulic lime) as the binder proved to be durable over many centuries and were used routinely until the late 1800s.[3] With the development of 'Roman' and then 'Portland' hydraulic cements, lime mortars became uneconomic from the builder's perspective as they set and harden slowly, and were perceived to be unsuited for use in wet situations. Lime mortars and renders were then superseded by cement gauged or pure cement mortars with their rapid hardening and high strength properties. Lime plasters were either gauged with Portland cement or gypsum, or entirely replaced by gypsum. As Portland cement was the dominant binder during the twentieth century, cement-rich mortars were used in the repair and restoration of historic buildings, where they had most often not been part of the original fabric.[4] This lack of compatibility between dense, impermeable cement-based repairs and traditional construction built in, and protected by, lime-based materials has resulted in masonry failure by cracking, water penetration and surface scaling.[5] In the last few decades there has been a resurgence in the use of lime due to recognition of its superior properties of breathability, flexibility and appearance.

Along with the lime revival has come an increased need to understand lime-based materials through laboratory investigation. When renovating or restoring a historic structure the conservation architect will often initiate a programme of materials' investigation with a view to the identification and formulation of appropriate repair materials. Consulting engineers and surveyors engaged in screening surveys of existing structures routinely request laboratory analysis to aid in the detection of decay, and diagnosis of the reasons for structural deterioration. When building new structures, specifiers and site engineers may request quality assurance testing of modern lime-based materials to check compliance with job specifications, sometimes as the result of unexpectedly poor performance, or following a perceived materials failure.

Although lime mortars, plasters and renders have been used for thousands of years, the laboratory techniques used to investigate them have only been developed relatively recently. Simple chemical analysis involving acid digestion has been utilized for nearly two centuries[6] while the petrographic microscope has been used to examine cementitious materials for over a century.[7] The application of sophisticated techniques such as scanning electron microscopy, instrumental chemical analysis and mercury porosimetry has only evolved over the last few decades.

The following article comprises a review of the techniques available to the materials investigator and discusses their relative advantages and disadvantages. Guidance is given regarding suitable on-site sampling techniques and interpretation of analytical results. A number of photomicrographs (photographs taken through the petrographic microscope) of mortars, plasters and renders have been included to illustrate features of interest. A brief technical introduction to lime technology is presented below.

LIME TECHNOLOGY OVERVIEW

Lime is formed by burning a source of calcium carbonate, usually lime-stone or magnesian limestone, between 850 and 1200°C, driving off carbon dioxide to form calcium oxide (quicklime). The calcium oxide is then slaked with water (evolving heat) to form calcium hydroxide (lime). The slaking process can produce dry lime hydrate powders or, if excess water is used, lime putty. Lime mortars, plasters and renders are produced by mixing, either quicklime, lime hydrate powder or lime putty, with an aggregate, and adding water if required. Lime made from pure lime-stone (non-hydraulic lime) sets by drying out, and then hardens wholly by absorption and slow reaction with carbon dioxide to become calcium carbonate once again.

Burning	$CaCO_3 + heat \rightarrow CaO + CO_2$
Slaking	$CaO + H_2O \rightarrow Ca(OH)_2$
Hardening	$Ca(OH)_2 + CO_2 \rightarrow CaCO_3 + H_2O$

When lime is burnt from limestone containing silica impurities it can develop partly cementitious properties, allowing it to set under water (hydraulic lime). This is due to the impurities in the limestone forming materials similar to those found in Portland cement, such as dicalcium silicate, aluminate and ferrite phases. The mechanism of hardening of hydraulic limes is a combination of carbonation (as above), and hydration of dicalcium silicate.[8] Portland cement is a highly processed form of arti-ficial hydraulic lime, which hardens by hydraulic reaction alone to form a binder chiefly comprising calcium silicate hydrate gel. Non-hydraulic limes can be gauged with Portland cement to impart a degree of hydraulic prop-erties. Calcium silicate hydrates are also formed when non-hydraulic lime with pozzolana additives react hydraulically. The formation of calcium silicate hydrates during the hardening of hydraulic limes, cements and pozzolana results in stronger, denser and less permeable mortars than when non-hydraulic lime is used alone as binder.

SAMPLING

The objectives of the investigation, and details of the proposed laboratory analysis, must be clearly defined before any sampling of mortar, plaster or render is attempted. A coordinated sampling programme should be prepared by professionals experienced with the repair of historic structures.[9] The number of samples required to achieve the objectives of the investigation will depend on the size of the structure, the types of construction used, and the number of construction phases (including repairs). The sampling regime should aim to ensure that the samples are as representative of the structure as possible, and the degree of sampling bias should be understood.

The actual sampling should preferably be performed by experienced operators who know what is required to achieve the objective of the investigation, and are familiar with the sampling demands of the prescribed analytical methods.

Samples may be carefully removed using hand tools (hammer and chisel) or by power tools (diamond drilling of core samples). Due consideration should be given to the health and safety issues that arise from the sampling activities, including safe access to the sample locations. In the case of plaster and render it is important to ensure that all of the coats are included, and the nature of the background material and the bond recorded. When sampling mortars from masonry walls it may be necessary to temporarily remove masonry units to ensure that original mortar is sampled, and the history of construction and past repairs understood. If intact lump samples cannot be obtained, powdered samples can be drilled out. However, although these may be suitable for chemical analysis, they are usually unacceptable for petrographic examination.

The samples should be placed immediately into separate sealed sample bags and each bag labelled with a unique identification reference. The exact sample locations should be recorded using a combination of written notes, drawings and photographs. These should include comprehensive inspection details of the location prior to sampling, and an 'as-found' record photograph. These details should be made available to the laboratory analyst if he/she is not present during the sampling operations.

The choice of analytical method determines the type of sample required. In order to avoid unnecessary damage to the structure, the size of the sample should be the minimum necessary to obtain accurate test results.[10] From the laboratory analyst's point of view the ideal sample would comprise one or more intact lumps with a total mass of around 200 g. This would allow a combination of petrographic examination and chemical analysis to be performed, while still retaining a small piece for reference purposes. However, when dealing with historic structures the 'ideal' sample may not be obtainable, and in such cases smaller samples are analysed on the understanding that they may be less reliable representations of the material from which they are removed. The overall representivity of samples may be improved by combining a number of small sub-samples to form one composite main sample.

METHODS

A plethora of analytical methods are available to today's materials investigators. The most important ones are listed below, with a brief explanation and discussion of their relative advantages and limitations. A guide to the selection of laboratory techniques for different investigation purposes is also provided, together with an indication of the relative costs – see Table 1. Techniques classified as 'essential' in the table are routinely used in commercial laboratories, while 'recommended' techniques are occasionally utilized. 'Rarely used' techniques are typically only used in academic research, or to meet particular specialized needs.

Visual and low-power microscopical examination of hand specimen

Visual examination is the essential first stage of any mortar analysis, and should start on-site with the aid of a hand lens (×10 magnification) to inspect the undisturbed material, and immediately after sampling. Once the sample is received in the laboratory this initial inspection is supplemented by examination at magnifications up to ×50 through a low-power binocular (stereoscopic) microscope. Combined with some simple physical and chemical tests this gives a rapid and economical provisional indication of the number of material types present; the number of coats; the grading, particle shape and mineralogy of the aggregate; the presence of inclusions, and the type and relative hydraulicity of the binder. It should be noted that definitive identification of the aggregate and binder composition requires more detailed high-power microscopical examination in thin-section (see below). The colour of both the aggregate and the binder should be classified by comparison with a Munsell™ colour chart.[11]

High-power microscopical examination of thin-section

Optical microscopy using the petrographic microscope at magnifications of up to ×600 is performed by specialist construction materials petrographers following guidance given in the American standard test method for examination and analysis of hardened masonry mortar, ASTM C1324.[12] The combination of visual/low-power examination in hand specimen,

Table 1 Selection of laboratory techniques for different investigation purposes.

Type of Investigation	Possible Information Required	Possible Analytical Methods		
		Essential Techniques	Recommended	Rarely Used
Conservation / Restoration / Archaeology / Quality Assurance:	Number of layers, their thickness and general appearance	V, P		
– Matching	Aggregate colour and particle shape	V	P	
– Raw material sources	Aggregate grading	V, P	G	
– Production techniques	Aggregate mineralogy	V, P		XRD, SEM
– Construction techniques	Binder colour	V		
– Compliance with specifications	Binder type & hydraulicity	V, P	C, XRD	SEM, XRF, AAS, ICP, T
	Presence of additives/impurities	IR		
	Mix proportions	C, P		
	Mix quality/workmanship	V, P		
	Porosity	P	MIP	
	Permeability	LP		
	Compressive strength	CS		
	Tensile strength	TS		
	Elastic modulus	EM		

Type of Investigation	Possible Information Required	Possible Analytical Methods		
		Essential Techniques	Recommended	Rarely Used
Screening for Deterioration and Failure Investigation:	Mortar hardness, friability	V		
– Structural integrity	Cracking/microcracking	V, P		
– Identifying causes of deterioration	Binder:aggregate bond	V		
	Bond to substrate	V, P		
	Leaching	V, P		
	Salt crystallization, salts content	V, P, H	XRF	AAS, ICP
	Freeze–thaw damage	V, P		

Microscopical Techniques:

V   Visual and low-power microscopical examination of hand specimen (£70)

P   High-power microscopical examination of thin-section (£175)

SEM   Scanning electron microscopy and electron microbeam analysis (£200/hour)

Chemical and Mineralogical Analysis:

C   Acid dissolution and wet chemical analysis for binder:aggregate ratio (£150)

G   Acid dissolution and sieve analysis for aggregate grading (£120)

H   Acid dissolution and wet chemical analysis for salts (£20–50)

XRD   X-ray diffraction analysis (£200)

AAS   Atomic absorption spectroscopy (£150)

XRF   X-ray fluorescence spectroscopy (£150)

ICP   Inductively coupled plasma atomic emission spectroscopy (£150)

IR   Infra-red analysis (£175)

T   Thermal analysis (£150)

Physical and Mechanical Testing:

MIP   Porosity by mercury intrusion porosimetry (£200)

LP   Laboratory permeability test (£200)

CS   Compressive strength of trial mix mortar (£150)

TS   Tensile strength of trial mix mortar (£150)

EM   Elastic modulus of trial mix mortar (£150)

Approximate 2012 costs in parentheses – stated as price per test except for SEM, which is charged at an hourly rate and more detailed high-power microscopical examination in thin-section, which are together called 'petrographic examination', has become the most useful all-round technique for the investigation of lime mortars, plasters and renders. A principal advantage of optical microscopy as an investigation tool is simply that one can see the evidence for oneself. A petrographic examination report will include colour photomicrographs to illustrate the general character and specific features of the material under investigation.

High-power microscopical examination requires the preparation of 'thin-section' specimens comprising 30 micron thick ground slices of the sample mounted on glass slides (typically 75 mm × 50 mm area). These are thin enough to allow light to pass for microscopical observation. This technique was originally developed by geologists for the study of rocks, and the thin-section preparation of lime-based mortars presents considerable challenges for the technician. As lime mortars are relatively soft and friable, it was not until the development of epoxy resins in the 1950s that these materials could be consolidated sufficiently to allow the production of thin-sections of the necessary quality.[13] Limes are also sensitive to heat and water, so that the drying and resin curing stages of specimen preparation have to be conducted at low temperatures (< 60°C), and coolants used during cutting and grinding must be based on oil and alcohol, rather than water.

Thin-section examination can provide a great deal of information regarding the ingredients used, their relative proportions, texture, porosity and condition. The aggregate for mortars, plasters and renders may be derived from different types of geological deposits, including natural sand resources (pit, river, beach, marine dredged) and rock formations (quarried). Natural sands often comprise rounded 'as-dug' particles, but may also include a proportion of more angular particles if oversize materials have been crushed down to the required size. Quarried rocks always have to be crushed down to achieve the required particle size/grading, and are consequently quite angular in shape. Occasionally, aggregates from more than one source are blended together to achieve desirable overall aggregate characteristics. The surface texture of aggregate particles, and the range of different particle sizes represented (grading), have a large influence on the quality of the resulting mortar. Desirable aggregates for modern brickwork mortars are usually uniformly graded with spherical, rounded particles to aid workability (Plate 1), whilst a repair specification for a historic render, in contrast, would require well graded aggregate with irregular, sub-angular particles to reduce drying shrinkage (Plate 2). Particle shape and grading characteristics are estimated through the microscope, sometimes by comparison with standard charts (Figures 1 and 2).

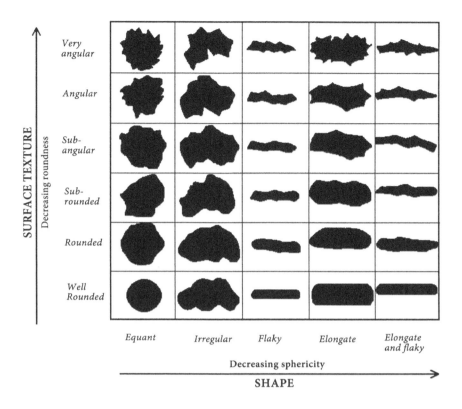

Figure 1   Classification of aggregate particle shape and surface texture.

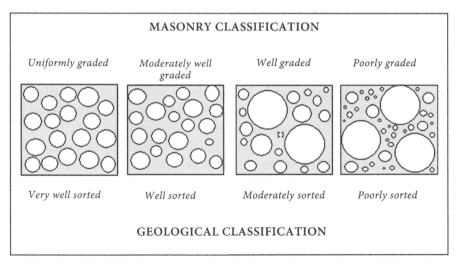

Figure 2   Classification of aggregate particle grading by visual assessment.

The mineralogy of the aggregate reflects its geological origin and gives clues to its source. For example, in south-east England the terrace sands of the River Thames were widely used (Plates 1 and 2); less widely used were crushed flints from the Cretaceous chalk (Plate 3). Aggregate may contain unacceptable amounts of potentially deleterious materials such as mica (Plate 4) or excessive fines, which may cause the mix to be weaker than intended; or pyrite, salts or organic matter, which may be unsound. Calcareous aggregates are used in areas where it is the most convenient source, for example in Oxfordshire, where the underlying geology comprises Jurassic limestone (Plate 5). At a Pembrokeshire lighthouse the shelly local beach sand was the easiest choice (Plate 6).

Lime binders have a distinctive texture, chiefly comprising finely crystalline calcite with occasional pockets of coarsely crystalline calcite deriving from the slow carbonation of trapped slaked lime.[14] The relative hydraulicity of the lime can be determined by observing the quantity of small relict grains chiefly comprising dicalcium silicate ('belite', Plate 7). Non-hydraulic limes have hardly any of these belite grains, while the number of grains observed increases progressively through the feebly hydraulic, moderately hydraulic and eminently hydraulic classes. The binders of historic lime mortars can contain anything up to 33% of lime inclusions.[15] Usually roughly spherical lumps of 50 microns to 10 mm in size, these inclusions typically consist of underburnt or overburnt quicklime (Plates 8 and 9). The presence of large (> 3 mm) unmixed lime lumps is sometimes interpreted to be caused by 'hot mixing' or 'dry slaking', a practice where damp aggregate is mixed with roughly crushed quicklime.[16] Underburnt or partially burnt inclusions often exhibit the original limestone texture, enabling identification of the geological source of the limestone burnt in the kiln (Plate 3).

If the lime has been gauged with Roman cement this will be detected through the observation of unhydrated cement grain relicts composed of dicalcium silicate crystals, and aluminous and ferrous interstitial phases. In the case of Portland cement, this will also include tricalcium silicate (Plate 10). These relict cement grains were relatively large (up to 500 microns across) in the early hydraulic cement mortars, and have decreased in size as clinker grinding technology has improved, with modern Portland cement mortars having cement grains of less than 30 microns in size. Lime plasters were sometimes gauged with gypsum, which is detected optically by the presence of finely disseminated gypsum and clay-rich lump impurities within the matrix, and large gypsum crystals lining air voids (Plate 11).

The advantages of adding pozzolanic materials to lime mortars have been known since antiquity, when the Romans included pozzolanic volcanic sand (pumice) in lime mortars for large-scale harbour structures around the Mediterranean.[17] Examination of these mortars shows clear evidence of pozzolanic reaction rims around the volcanic sand particles (Plate 12). Crushed ceramic tiles and potsherds were used in Roman mortars for small-scale and terrestrial purposes (Plate 13).[18] Brick dust has been added to English lime renders since the eighteenth century making them dark brown in colour (Plate 14). The degree of pozzolanic activity of ceramic additives depends on factors such as the firing temperature of the clay, and the particle size of the pozzolana, which can be investigated microscopically. The presence of other pozzolanas such as pulverized-fuel ash (pfa), ground granulated blastfurnace slag (ggbs) and trass are easily detected optically due to their distinctive morphologies, while very finely divided mineral additions such as metakaolin and microsilica (silica fume) are too small to be seen. The detection of very finely divided mineral additives relies on observation of indirect evidence such as high optical density, and portlandite depletion of the binder, and in the case of microsilica, the presence of ball-shaped agglomerations of undispersed material.[19]

Animal hair and straw have historically been used as reinforcement for lime plasters (Plate 15). Impurities such as charcoal contamination from the kiln fuel (Plate 16) or vegetable matter (Plate 17) are often seen through the microscope.

The volume of air voids, their size and shape are easily observed in thin-section. The quantity of large (> 1 mm diameter), irregular shaped entrapped air voids is estimated to assess the effectiveness of compaction at the placing. Small (< 1 mm diameter), spherical air voids are entrained in the mix for a number of reasons, and an abundance of such air voids may indicate the use of air entraining (surface–active) or plasticizing (water-reducing) admixtures.

Microscopical examination is eminently suited to screening for evidence of decay caused by leaching, salt attack, freeze–thaw damage and, for mortars including Portland cement, sulfate attack. Signs of matrix replacement or recrystallization, the presence of secondary deposits and the presence of filled or unfilled cracks and micro-cracks (Plate 18) may provide evidence of deterioration caused by deleterious chemical reactions.

An estimation of the quantity of calcareous aggregate and lime lumps can be made through the microscope, and used to correct possibly erroneous assumptions about mortar mixes which can arise from the inherent inaccuracies of acid dissolution and wet chemical analysis (see below). Mix proportions can also be determined directly from the thin-section by three methods:

1. By visual comparison with reference to thin-sections of mortar of known mix proportions.[20]
2. By determination of the area occupied by each mortar constituent (aggregate, lime, lime lumps and air voids) through point counting, and then calculating the mix proportions using assumed densities.[21]
3. A variation of method 2 where areas occupied by the various mortar constituents is determined by digital image analysis.[22]

Method 1 is performed in just a few minutes, but is the least accurate of the three. Method 2 can give very accurate results, but point counting of the thin-section is labour-intensive and time-consuming. Method 3 can produce very accurate results rapidly; however, few laboratories are currently equipped with the necessary image analysis software. For all three methods care should be taken to ensure that the plane of thin-section is as representative of the whole material as possible.

Acid dissolution and wet chemical analysis

The mix proportions of lime mortar samples are commonly determined by simple chemical analysis in accordance with BS 4551.[23] A representative sample is powdered, and hydrochloric acid used to dissolve away the carbonate binder and soluble silica, leaving the insoluble aggregate residue. Determined values of calcium oxide, soluble silica and insoluble residue allow the binder:aggregate ratio and the degree of binder hydraulicity to be estimated. However, this technique has a number of limitations, the main one being that carbonate aggregate particles (limestone, marble, shell) will be dissolved away with the binder, thus giving an incorrectly high assessment of binder content. Values of soluble silica used for the estimation of the degree of hydraulicity of the lime binder are also subject to error if pozzolans are present, due to the presence of soluble silica reaction rims around pozzolanic particles.

Acid dissolution and sieve analysis

British Standard BS 4551 also includes a method for the determination of aggregate grading by sieve analysis. This involves dissolving away the calcium carbonate binder from the mortar using hydrochloric acid, and passing the dried insoluble residue through a set of test sieves of decreasing aperture. By recording the mass retained on each of the sieves, a cumulative curve of particle size distribution can be plotted. As above, a major disadvantage of this method is that any carbonate aggregate particles will be dissolved away and therefore not recorded in the results. Care must also be exercised when attempting to match the colour of the aggregate residue, as acid digestion may have altered the colour. For the purposes of most investigations obtaining aggregate grading and colour data by petrographic examination is preferable.

Wet chemical analysis of salts content

Determination of salt content is typically required to investigate the causes and severity of decay. High salt levels can indicate the presence of decay, the exact cause of which would normally be further investigated by petrographic examination. High levels of sulfate, chloride and nitrate salts can cause decay of porous building materials by salt crystallization. High levels of sulfate can result from the leaching of adjacent clay brick masonry or soil to cause the formation of expansive gypsum (calcium sulfate hydrate) within lime mortar (Plate 18). Lime mortars gauged with Portland cement are susceptible to sulfate attack, a reaction where calcium aluminate phases of the Portland cement are converted to expansive ettringite (calcium aluminium sulfate hydrate) by sulfate bearing solutions (Plate 19). Lime mortars gauged with Portland cement and including limestone aggregate are potentially susceptible to the thaumasite form of sulfate attack (TSA), a reaction resulting in replacement of cement hydrates by thaumasite (calcium carbonate silicate sulfate hydroxide hydrate), which requires a source of carbonate and sulfates, as well exposure to cold wet conditions (Plate 20).[24] Determination of hygroscopic salt content (chlorides and nitrates) is useful when investigating damp problems in structures. High levels of hygroscopic salts combined with other evidence can indicate that the moisture has originated from the ground, rather than as a result of condensation or rainwater penetration. Wet chemical analysis, in accordance with British Standards, is the most economical method of accurately determining salt contents.[25] More expensive analytical techniques such as X-ray fluorescence are also suitable for quantitative determinations.

Instrumental chemical and mineralogical analysis

Sometimes more sophisticated instrumental methods of chemical analysis are required. Many of these techniques can be performed on very small samples (a few grams or less). Of the various types of instrumental analysis, spectroscopic techniques are the most successful when applied to lime-based building materials. X-ray diffraction (XRD) provides mineralogical determination of crystalline phases of the binder, including pozzolans, hydraulic reaction products and mineral constituents of the aggregate.[26] XRD has the advantage of actually identifying the minerals present, although estimation of the relative amounts is at best only semi-quantitative. Other spectroscopic methods such as X-ray fluorescence (XRF), atomic absorption (AAS) and atomic emission (ICP-AES) give very precise determinations of elemental composition. However, the elemental composition data from these techniques requires skilled interpretation by the analyst in order to identify the proportions of mineral phases present. These techniques will only detect inorganic compounds: to identify the presence of organic additives such as tallow or linseed oil, infrared spectroscopy is the method of choice. Thermal analysis, a range of methods which record the temperature at which mortar components decompose, is occasionally also used. Its most successful application has been for determining the binder type, but the results should not be used in isolation.[27]

Scanning electron microscopy

The scanning electron microscope (SEM) enables very small structures to be imaged (down to a fraction of a micron) allowing examination of the morphology of lime mortar components and their textures. The imager can be used to locate targets for accurate elemental analysis using the SEM's on-board electron microprobe (EDS or EDX). The electron beam can be focused on specific features observed by imaging for precise identification, or scanned over an area of several square millimetres to create a map showing the relative abundance of a range of different elements across the target area.

Physical and mechanical testing

Physical parameters of lime-based construction materials, such as porosity and permeability, can have a great influence on their performance. The sizes and relative abundance of pores within mortar can be determined by mercury intrusion porosimetry; with the use of specialized software pore structures can be visualized in three-dimensional models, and the crystal size distribution of the binder can be calculated.[28] Water vapour permeability can be assessed using desiccants as driving agents, and gas permeability can be measured by nitrogen injection techniques.[29,30] Durability assessment by accelerated exposure to freeze–thaw and salt crystallization mechanisms has been performed on mortars, although standardized methods have yet to be published.[31]

Physical properties for existing lime materials in historic buildings, such as compressive strength, tensile strength and modulus of elasticity, are usually not measured directly due to the difficulty of obtaining sufficiently large samples. If required, it is possible to test these properties by duplicating trial mixes of new mortars by casting cube or cylinder specimens of the required dimensions. The strength of mortars in existing joints is normally estimated by comparing chemically determined mix proportion results to the strengths of modern mortars of similar compositions. Where pure lime mortars have been identified in an old building the mortar strength will often be assumed to be around 0.5–1.0 N/mm², although this approach usually underestimates the actual compressive strength of historic mortars.[32] An indirect test giving an indication of the compressive strength of masonry mortars, the 'screw pull-out test' has been developed for use on-site.[33]

INTERPRETATION OF RESULTS

When matching lime-based mortars for restoration work, the aim is usually to produce a material which is compatible first with the type and condition of the masonry, and second to the particular circumstances of exposure. The choice is primarily based on knowledge of the properties of various mortars, and is not arrived at by analysis of the mortar alone.[34] Whether or not a repair mortar is compatible will depend primarily on its appearance, strength, flexibility and permeability. These properties are in turn determined mainly by the choice of aggregate, binder type and their relative proportions and working characteristics.

Historically, aggregates for mortars used in vernacular building were usually obtained from within 50 miles of the site due to access and transport costs.[35] Data from petrographic examination can be used to pinpoint the original source by comparing the petrographic data with geological and historical information. The original source may no longer be operational, and it may be impossible or inappropriate to commence new extractive works due to planning restrictions. In these circumstances a good match may be found by comparison with that of commercially available material. This is a visual process, supplemented by petrographic examination findings, with the main aggregate characteristics considered being colour (often controlled by mineralogy), grading and particle shape.

Care must be taken when interpreting the results of wet chemical analysis to ensure that the presence of calcareous aggregate has not led to an overestimation of the binder content. Chemical analysis of historic mortars often generate results suggesting very binder-rich mixes (1:1 or 1:2, lime:sand) which contrasts with the 1:3 (lime:sand) ratio commonly specified in current building work. This apparent discrepancy may well be due to the chemical analysis interpreting lime inclusions (unmixed lime, underburnt or hardburnt particles) commonly found within historic lime mortar as part of the binder. Although derived from the lime, these inclusions do not function as binder in the mortar, and in terms of performance can be regarded as a form of aggregate.[36] However, for the purposes of matching, it is the original binder content including the lime inclusions which is used, so that the repair lime should ideally include similar types and amounts of inclusions. In addition to identification of the binder type and its relative hydraulicity, a description of the nature and quantity of calcareous aggregate and lime inclusions is provided by petrographic examination. The combination of petrographic examination and wet chemical analysis is a powerful tool for the investigation of historic lime mortars, plasters and renders.

To aid the process of locating aggregate and lime sources, lists of suppliers and test data for some currently available building sands and lime products suitable for use in restoration have been published.[37,38] It should be noted when using the Directory of Building Sands and Aggregates that only 60% of the sources in England are listed. Addresses of other potential sources in the United Kingdom can be found in the Directory of Mines and Quarries or the Directory of Quarries and Quarry Equipment.[39,40] Information regarding Scottish aggregate sources can be found in Scottish Aggregates for Building Conservation.[41]

CONCLUSIONS

When selecting a laboratory to carry out investigations of lime-based building materials it is best to choose an organization that offers a wide range of investigative techniques. This way the analyst is likely to recommend the correct technique for specific requirements of the job, rather than one simply from a limited range to which they have access. Many laboratories are accredited by the United Kingdom Accreditation Service (UKAS) which gives some reassurance of the quality of the laboratory work undertaken.

From the myriad of analytical techniques available the ones most commonly used are:

- visual & low-power microscopical examination of hand specimen;
- acid dissolution and wet chemical analysis;
- high-power microscopical examination of thin-section.

Wet chemical analysis is subject to a range of inaccuracies and uncertainties and should only be used in isolation to check mix proportions of new mortars where the ingredients used are known. When matching historic mortars for restoration works the common practice is to subject each sample to a combination of visual/low-power microscopical examination and wet chemical analysis to give an indication of the likely ingredients and properties, from which a combined judgement can be taken. If the petrographer finds evidence of calcareous aggregate during the visual examination, lime lump inclusions, distress or deterioration, it is essential that high-power microscopical examination of thin-section is performed to calibrate and correct the wet chemical analysis results.

Sometimes scanning electron microscopy or X-ray diffraction is necessary to crosscheck and clarify the findings of the high-power microscopical examination. Infra-red analysis is occasionally used when details of organic additives are required.

For plates referred to in this paper, see the colour section following page 132.

This paper was first published in the *Journal of the Building Limes Forum*, Volume 10, 2003. This is a revised and updated version.

Notes

1 Historic Scotland, *Mortars in Historic Buildings: A Review of the Conservation, Technical and Scientific Literature*, Historic Scotland, Edinburgh, 2003.
2 Ibid.
3 Oates, J. A. H., *Lime and Limestone: Chemistry and Technology, Production and Uses*, Wiley-VCH, Germany, 1998.
4 Historic Scotland, *op. cit.*

 5 Ashurst, J., *Mortars, Plasters and Renders in Conservation* (2nd edition), Ecclesiastical Architects and Surveyors Association, London, 2002.

 6 Ibid.

 7 St John, D. A., Poole, A. B. and Sims, I., *Concrete Petrography, a Handbook of Investigative Techniques*, Edward Arnold, London, 1997.

 8 Ibid.

 9 British Standards Institution (BSI), *BS EN 8221-2 Code of practice for cleaning and surface repair of buildings – Part 2: Surface repair of natural stones, brick and terracotta*, BSI, London, 2000.

10 Historic Scotland, *op. cit.*

11 *Munsell™ Soil Color Charts, Revised Edition*, Macbeth Division of Kallmorgan Instruments Corporation, USA, 1994.

12 ASTM C1324-10, *Standard test method for examination and analysis of hardened masonry mortar*, ASTM International, Philadelphia, USA, 2010.

13 St John et al., *op. cit.*

14 Ibid.

15 Leslie, A. B. and Hughes, J. J., 'Binder microstructure in lime mortars: implications for the interpretation of analysis results', *Quarterly Journal of Engineering Geology and Hydrology*, Vol. 35, 2002, pp. 257–63.

16 Hughes, J. J. and Leslie, A. B., 'The petrography of lime inclusions in historic lime based mortars', *Proceedings of the 8th Euroseminar on Microscopy Applied to Building Materials*, Athens, 2001.

17 Brandon, C., 'Caesarea Papers 2 – Pozzolana, lime and single-mission barges (Area K)', *Journal of Roman Archaeology*, Supplementary Series Number 35, 1999, pp. 168–78.

18 Siddall, R., 'The use of volcaniclastic material in Roman hydraulic concretes: a brief review', *The Archaeology of Geological Catastrophes*, eds. McGuire, W. G., Griffiths, D. R., Hancock, P. L., and Stewart, I. S., Geological Society, London, Special Publications Vol. 171, 2000, pp. 339–44.

19 St John et al., *op. cit.*

20 Palmer, J. J., 'Whose lime is it anyway?', *Natural Stone Specialist*, Vol. 33, 1998, pp. 29–32.

21 'Rilem TC 167-Con: Characterisation of old mortars', RILEM Draft Recommendation, *Materials and Structures*, Vol. 34, 2001, pp. 387–88.

22 Mueller, U. and Hansen, E. F., 'Use of digital image analysis in conservation of building materials', *Proceedings of the 8th Euroseminar on Microscopy Applied to Building Materials*, Athens, 2001.

23 British Standards Institution (BSI), *BS 4551-2005 Mortar. Methods of test for mortar. Chemical analysis and physical testing*, BSI, London, 2010.

24 DETR, 'The thaumasite form of sulfate attack: Risks, diagnosis, remedial works and guidance on new construction', *Report of the Thaumasite Expert Group*, Department of the Environment, Transport and the Regions, London, 1999.

25 BS 4551, *op. cit.*

26 Groot, C. J. W., Bartos, P. J. M. and Hughes, J. J., 'Historic mortars: Characteristic and tests – Concluding summary and state-of-the-art', *Proceedings of RILEM International Workshop PRO12*, Paisley, 2000, pp. 443–54.

27 Ellis, P. R., 'Analysis of mortars (to include historic mortars) by differential thermal analysis', *Proceedings of RILEM International Workshop PRO12*, Paisley, 2000, pp. 133–47.

28 Middendorf, B., 'Physio-mechanical and microstructural characteristic of historic and restoration mortars based on gypsum: current knowledge and perspective', *Natural Stone, Weathering Phenomena, Conservation Strategies and Case Studies*, eds. Siegesmund, S., Weiss, T. and Vollbrecht, A., Geological Society, London, Special Publications 205, 2002, pp. 165–76.

29 Banfill, P. F. G. and Forster, A. M., 'A relationship between hydraulicity and permeability of hydraulic lime', *Proceedings of RILEM International Workshop PRO12*, Paisley, 2000, pp. 173–83.

30 Valek, J., Hughes, J. J. and Bartos, P. J. M., 'Portable probe gas permeability in the testing of historic masonry and mortars,' *Proceedings of RILEM International Workshop PRO12*, 2000, pp. 185–96.

31 Groot et al., *op. cit.*

32 The Institution of Structural Engineers, *Appraisal of existing structures*, (2nd edition), SETO Ltd, London, 1996.

33 BRE, 'Measuring the compressive strength of masonry materials: the screw pull-out test', *BRE Digest 421*, Building Research Establishment, Watford, 1997.

34 Ashurst, J. and Dimes, F. G., *Conservation of Building and Decorative Stone*, Butterworth Heinemann, Oxford, 1998.

35 English Heritage, *The English Heritage Directory of Building Sands and Aggregates*, Donhead Publishing Ltd, Shaftesbury, 2000.

36 Leslie, *op. cit.*

37 English Heritage, *op. cit.*, 2000.

38 English Heritage, *The English Heritage Directory of Building Limes*, Donhead Publishing Ltd, Shaftesbury, 1997.

39 Cameron, D. G., *Directory of Mines and Quarries*, British Geological Survey, Keyworth, 2008.

40 Quarry Management, *Directory of Quarries & Quarry Equipment* (32nd edition), QMJ Publishing, Nottingham, 2009/2010.

41 Historic Scotland, *Technical Advice Note 19 – Scottish Aggregates for Building Conservation*, Historic Scotland, Edinburgh, 1999.

Diagnosing Defects in Lime-Based Construction Materials

Jeremy Ingham

INTRODUCTION

It is widely understood that lime-based construction materials are frequently encountered in historic structures around the world, including domestic dwellings, stately homes, public and commercial buildings and ecclesiastical buildings. It is less well recognized that lime mortars were utilized for vast numbers of other civil engineering structures such as bridges, roads, canals and sewers. For example, in the UK, lime mortars are present in many of the national stock of around 40,000 masonry arch bridges.[1] Consequently, knowledge of lime technology and associated defects is crucial to understanding the condition and conservation requirements of historic buildings and civil infrastructure. It is also required to ensure that new build structures using lime are constructed correctly, and to ensure their serviceability, as in the last few decades there has been a modest resurgence in the use of lime for new build.

This paper describes the techniques available to ensure correct diagnosis of defects in lime mortars, plasters and renders. It discusses the various types of defects, their causes, and the techniques used for their investigation. It also describes typical repair options. This paper concentrates on site investigation techniques, as detailed descriptions of laboratory investigation techniques are available elsewhere.[2,3,4]

Figure 1 Deterioration of a guard tower on an unrestored section of the Great Wall of China at Panjiakou Reservoir, Hubei.

TYPES OF DEFECTS AND THEIR CAUSES

Over time, exposure to natural weathering processes and in-service wear and tear cause deterioration of lime-based construction materials. This may be exacerbated by poor design, the use of inappropriate materials or inadequate workmanship.

Deterioration processes can result in a range of defects including cracking, debonding, loss of strength and surface erosion. Lime mortars, plasters and renders are designed to be weaker than masonry units (i.e. stone or clay bricks) and are intended to decay in a sacrificial manner in order to protect these units. Consequently, historic masonry structures require maintenance and repair during their service life, to repoint eroded mortar joints and reinstate protective render and plaster. Figure 1 shows an unrestored section of the Great Wall of China that exhibits the effects of deterioration, resulting from a prolonged absence of maintenance. Determining the causes of deterioration and defects is necessary in order to correctly specify repairs. Failure to understand the causes of deterioration can lead to early re-occurrence of defects, or even accelerated decay of the masonry.

Table 1 provides a summary of the types and typical causes of defects

Table 1 Summary of defects affecting lime-based construction materials and their causes.

Defects	Manifestation	Possible causes
Structural cracking	Cracks greater than 1 mm width that extend into substrate	Structural movement Background movement Impact Thermal movement Severe weathering
Shrinkage cracking	Cracks greater than 0.5 mm width that are confined to the render/plaster/ mortar	Too much binder Unsuitable aggregate Insufficient hair content Oversaturated background Presence of stud lines in substrate
Crazing	Pattern of hair-line cracks on plaster/render surface	Over trowelling Inadequate curing Too much binder High fines content in aggregate
Surface cracking, spalling, discolouration	Explosive surface loss, reddening	Fire damage
Loss of bond	Hollow areas, absent, delamination or bulging of plaster/render	Poor background preparation Background movement Render stronger than background Render too thick Poor suction control Leaching Freeze–thaw action Salt crystallization Impact
Inadequate strength	Weakness, friability, powdering of surface	Unsuitable material choice for exposure conditions Insufficient binder content Uneven mixing Inadequate compaction Poor suction control Inadequate curing Leaching Salt crystallization Sulphate action Freeze–thaw action
Paint coating failure	Blistering/ peeling/flaking paint coating	Leaching behind paint Unsuitable impermeable paint used Limewash due for renewal
Staining	Green, white or black staining	Organic growth Pollution coating Leaching deposits Salt efflorescence

in lime-based construction materials. In broad terms, defects are caused by one or more of the following factors:

- inappropriate design/specification;
- poor quality materials;
- inadequate workmanship;
- weathering and decay mechanisms;
- deleterious reactions.

Both the original design and design of any alterations will influence the in-service performance of lime mortars, plasters and renders. If the structural design is inadequate, movement of the structure can cause cracking of mortars and debonding of plasters and renders. In addition, the design detailing will help to define the weathering micro-environments on both the external and internal elevations. For example, provision of details to shed water from roofs away from the walls below will slow the weathering of external elevations. Also, a design that provides good weatherproofing will protect the interior elevations from water penetration.

Mortars

If mortars that are stronger, less permeable or less flexible than the bricks or stones in a wall are specified, deterioration will occur preferentially to the masonry units. Sympathetic specification of materials is of particular importance when conducting repairs of historic structures.

In Britain, Portland cement-rich mortars have been used for repair and restoration of many historic buildings, where they have most often not been part of the original fabric. This has resulted in a lack of compatibility between dense, impermeable cement-based repairs and traditional lime-based construction, causing masonry failure by cracking, water penetration and surface scaling. Figure 2 shows a fourteenth-century church constructed from 'clunch' stone (a type of hard chalk) and lime mortar. Repairs using Portland cement resulted in a considerable amount of falling pieces of masonry. Figure 3 shows an example of a flint wall that has been incorrectly constructed using Portland cement instead of hydraulic lime mortar. The wall suffered early failure as the strong, impermeable and shrinkage-prone cement was unable to adequately withstand weathering.

Figure 2 Masonry falling from a church spire caused by inappropriate Portland cement repairs to clunch stone masonry. Ashwell, England.

Figure 3 Early failure of a flint wall owing to the inappropriate use of Portland cement mortar to bed the flints. Hertfordshire, England.

Figure 4   Early failure of impermeable paint coating on lime render of an eighteenth-century building façade. St Albans, England.

Inappropriate materials

Wall surfaces have long been painted to improve their appearance and durability. Traditional coatings such as limewash are often relatively permeable, and thus able to protect the wall while allowing it to breathe and dry out.

Synthetic paints with plastic-based binders were introduced in the 1930s. The substitution of such synthetic paints for earlier limewashes can cause deterioration, as many do not allow dampness trapped within the masonry to dry out. This can be exacerbated by the accumulation of salts behind the coating. Figure 4 shows a modern impermeable paint on the lime-rendered exterior of an eighteenth-century building façade detaching from the render. Site inspection revealed that moisture penetrating a cracked coping was migrating into the render and becoming trapped behind the impermeable paint, causing it to debond.

Poor quality ingredients will detrimentally affect lime mortars. Lime binders that contain an excessive proportion of underburnt or overburnt lime lumps will be of inferior strength.

A number of defects, such as shrinkage cracking, crazing and loss of bond, are associated with the use of inappropriate aggregates. Those for use with lime binders should be well graded with irregular, sub-angular particles ('sharp sand') and minimal fines. Poorly graded 'soft sand' with rounded particles is specified for internal plaster in order to increase workability.

If required, hair for lime plasters and renders should be added in sufficient quantity (4–8 kg/m³), evenly distributed, strong, soft, not too springy and of various lengths between 25 mm and 100 mm. Goat and cattle hair are the most suitable as they are coarse and covered with tiny barbs that hold hairs in place within the lime binder. Horse hair has a smooth surface and can consequently be pulled away from the hardened mortar by hand. Most human hair is considered too fine and of insufficient strength.

The addition of impurities to a mix can sometimes cause problems. For example, if large quantities of charcoal or furnace clinker are added to colour the mortar black, the sulfates they contain could cause a deleterious reaction.

Workmanship

It can be difficult to find a contractor who is skilled at lime work, and inadequate workmanship is consequently a common cause of defects. Failure to prepare the background adequately, or overwatering of the substrate can both inhibit bonding. Failure to control suction of the background will cause rapid drying resulting in weak plasters and renders. Failure to provide a stable background leads to cracking and debonding. Figure 5 shows an example of a traditional cottage that was completely re-rendered following subsidence and underpinning. The contractor used an expanded metal mesh as the background, rather than the traditional timber lath that had been specified. The mesh was poorly fixed and unstable, resulting in cracking due to background movement. The render was also intrinsically very weak, and when examined through the polarizing microscope showed a network of microscopic tension gashes caused by rapid drying.

Inaccuracies in batching can cause low strength from too little binder, or shrinkage cracking and crazing from an excess. Overwatering the mix (adding too much water during mixing) will also lower the strength and make the material more prone to drying shrinkage problems. Failure to adequately disperse the ingredients during mixing is another potential cause of low strength, as is under-compaction during placing, which leaves air voids.

Figure 5   Breakout of render on a historic cottage to reveal the background. Note vertical crack to left. Suffolk, England.

If plaster or render layers are applied too thickly, they are liable to sag through their own self-weight, and may debond as a result. Thick layers of non-hydraulic lime mortar will also take longer to carbonate and gain strength and physical hardness. The hardening of lime by carbonation is a delicate process and lime work must be protected both from rapid drying from sun and wind, as well as from driving rain and frost as it is curing.

Weathering and impact damage

Natural weathering mechanisms will eventually cause the deterioration of all masonry elements. The main decay mechanisms are thermal movement, rain washing, frost action, salt crystallization and biodeterioration. Different parts of all structures will have different micro-environments that will dictate which decay process will be dominant. On areas of buildings which can remain wet (plinths, sills, string courses and copings), frost action can be a major cause of surface erosion and cracking. Thought to be related to a build-up of expansive forces, the mechanisms of frost damage to masonry are complicated and not yet entirely understood. The pore structure of masonry materials plays an important part in determining resistance to frost, with free-draining materials being more durable.

Figure 6 Heavy leaching along mortar joints of a brickwork retaining wall. Leached deposits of calcium carbonate appear white.

Salt crystallization involves the expansive growth of crystals disrupting wall surfaces. The salts may originate from internal sources such as salt-rich masonry units or the mortar or from external sources such a groundwater or sea spray. Moisture movements through masonry can dissolve away the binder within mortar with consequent loss of strength (Figure 6).

In addition to natural weathering, structures are subject to physical wear and tear from use during their service life. Damage also results from specific impact events such as vehicle strikes. In certain circumstances there may be a requirement to conserve the damage if it has some historical significance. For example, impact damage from bullets and shrapnel sustained during wartime often require retention.

Biodeterioration and pollution

Certain organisms that colonize the built environment have the potential to cause biodeterioration in lime mortars. Trees, creepers and mosses cause physical damage from their roots (Figure 1), while certain bacteria, fungi and lichens deposit organic acids that dissolve the carbonated lime binder. Corrosive acids released from bird droppings also dissolve lime. In addition, the presence of organic growth allows the surface to remain damp for longer, promoting the action of leaching and frost weathering.

The most common deleterious chemical reactions that effect lime-based construction materials are sulfate action and acid attack. High levels of sulfates can result from the sulfate-rich inclusions in mortar, adjacent clay brick masonry or soil, or polluted masonry surfaces, to cause the formation of expansive gypsum crystals (calcium sulfate hydrate). This expansive growth of gypsum and other salts can cause cracking and other deterioration. In industrialized parts of the world, pollution from industry and transport causes masonry decay by attack from acid gases in the air. Sulfur and nitrogen-based gases react with moisture to produce acidic solutions that can directly attack calcareous materials such as lime mortars and renders. Exposed masonry surfaces are dissolved away and salt crusts often including particulates and other dirt are deposited on more sheltered surfaces, giving a soiled appearance.

Fire

Historic buildings can be seriously affected by fires, and lime mortars can suffer considerable damage when exposed to high levels of heat. Fire causes surface cracking, spalling and calcination of the lime binder, all of which can weaken lime mortars. Significant thermal damage starts at around 300°C and calcination of calcium carbonate binder occurs at > 600°C. A load-bearing masonry wall exposed to fire will suffer a progressive reduction in strength due to deterioration of the mortar.[5] In addition to structural damage, heating may cause irreversible discolouration that can be a serious aesthetic issue. For example, iron compounds present within some mortar aggregate particles will oxidize in the range 250–300°C and turn red.

INVESTIGATION TECHNIQUES

Correct diagnosis of defects will often require a carefully planned investigation involving a range of techniques. Planning an investigation may involve a desk study of available drawings, specifications and other records, often supplemented by a visit to the site to conduct a preliminary survey. The locations for invasive surveys and sampling of materials clearly need to be established. Aspects which also need to be considered in the investigation plan include providing safe access; protecting and minimizing disruption to users of the structure; defining the area for visual survey, and minimizing the damage to the structure that could be caused by the investigation. Methods for sympathetic reinstatement of sampling locations need also to be prepared and applied.

On any valuable building it may be necessary to obtain consent before undertaking an invasive survey. In the UK, for instance, special consents are required before intervening in any existing structures that falls within the following categories (this is not an exhaustive list):

- a scheduled ancient monument:
- listed building:
- an unlisted building in a conservation area:
- a building belonging to the Crown;
- a Church of England building used for worship;
- a resting place for bats.

One must also obtain the owner's and/or tenant's authority to enter the structure, and inform any other occupiers about the purpose of the visit.

Investigation techniques can be divided into those that are conducted on-site and those conducted off-site in specialist laboratories. The various techniques are summarized in Table 2.

On-site, the structure will typically be subjected to a detailed visual inspection. Ideally, inspection should be conducted from touching distance, and provisions to allow access, such as scaffolding, elevating platforms or industrial rope access may be required. If this is not possible, a lower quality of visual survey can be conducted from ground level using binoculars, although this is likely to be significantly less definitive. For renders and plasters it is useful to conduct a tapping survey (using a small hammer or metal rod) to detect hollow areas that have debonded from the substrate, but have not yet fallen from the wall.

It may sometimes be appropriate to remove all loose material to prevent injuries from falling debris (subject to obtaining appropriate permissions) or, alternatively, to restrict access to the structure until proper repairs can be carried out. The findings of the visual survey are typically recorded on detailed drawings, often using hand-held computers, supplemented by photographs or even video. Figure 7 shows an example of a record drawing produced during a render condition survey of an eighteenth-century building façade in England.

Table 2 Typical on-site and laboratory investigation programmes for diagnosing defects in lime-based construction materials.

Defect Type	Investigation Programme	
	On-site investigation	Laboratory testing
Structural cracking	S1, S2, S3, S4	L1, L2, L3 & L4
Other cracking	S1, S2, S3, S4	L1, L2, L3 & L4
Surface crazing of plaster or render	S1, S3, S4	L1, L2, L3 & L4
Debonded/detached plaster or render	S1, S3, S4	L1, L2, L3 & L4
Low strength, surface scaling/friability of mortar, plaster or render	S1, S3, S4	L1, L2, L3 & L4
Fire damage	S1, S3, S4	L1, L2, L3, L4
Coating failure	S1, S3, S4, S5	L1, L2, L3, L4, L5, L6
Staining	S1, S3, S4, S6	L1, L2, L3, L4, L5

S1	Visual inspection & tapping survey
S2	Crack mapping and crack width measurement
S3	Breakout along crack to background and/or boroscope survey
S4	Sampling of mortar/plaster/render from defective areas and at least one non-defective area
S5	Sampling of coating from defective areas and at least one non-defective area
S6	Sampling of staining/soiling deposits
L1	Visual/low-power microscopical examination of all samples
L2	High-power microscopical examination of selected samples
L3	Mix proportions determination of all coats of selected samples
L4	Sulfate content analysis
L5	Salt content analysis
L6	Coating analysis

INVASIVE INVESTIGATION

It may be appropriate to conduct an invasive investigation to determine the nature of materials present, allow boroscope inspection and/or to obtain samples for laboratory testing. The number of locations which need to be opened up to achieve the objectives of the investigation will depend

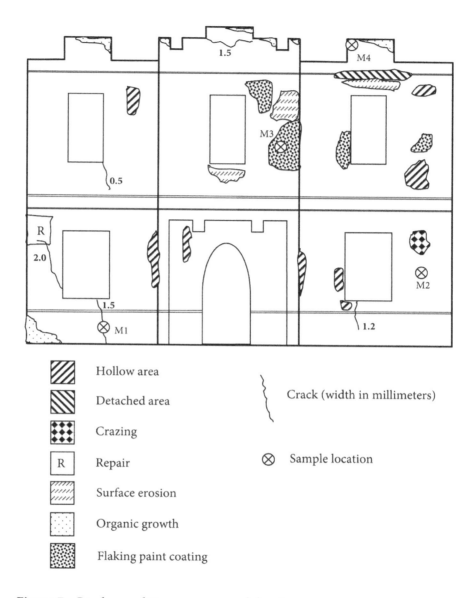

Figure 7   Render condition survey record drawing.

on the size of the structure, the types of construction used and the number of construction phases (including past repairs).

Boroscope inspections are normally conducted through small (10–16 mm diameter) drilled holes. Breakouts should be carefully undertaken using hand tools such as hammer and chisel, or by power tools, for example diamond drilling of core samples. Due consideration should be given to the health and safety issues that arise from such activities. In the case of plaster and render it is important to ensure that all of the coats are investigated and the nature of the background material and the bond recorded. The breakout shown in Figure 5 was undertaken to investigate the causes of cracking and weakness of new lime render on a historic cottage. The breakout revealed the presence of an inappropriate and poorly installed expanded metal lath background. The breakout also allowed convenient sampling of all three layers of render for laboratory testing.

When investigating mortars from masonry walls, it may be necessary to temporarily remove masonry units to ensure that the mortar being sampled is original, and the full history of construction and repairs understood. The smallest useful sample for laboratory testing is 20 g weight of intact material. It is important to ensure that the sample is representative, and it may be appropriate to extract samples of up to 300 g weight for extensive laboratory testing regimes. The samples should be placed immediately into separate sealed sample bags and each bag labelled with a unique identification reference. The exact sample locations should be recorded using a combination of written notes, drawings and photography. These details should be made available to the laboratory analysts if they are not present during the sampling operations.

Laboratory tests

Over the last 40 years a number of laboratory tests suitable for the investigation of lime-based construction materials have been developed.[6] Laboratory tests are primarily used to determine the ingredients present, the manufacturing techniques and the mix proportions in order to enable a matching material to be specified for repairs. However, laboratory tests can also be extremely helpful in determining the causes of failures by providing insights into the quality of workmanship, and by screening for evidence of distress or deterioration.

The most commonly used laboratory test is 'wet' chemical analysis of mix proportions to BS 4550[7] or ASTM C1324.[8] This test is subject to significant errors, particularly if calcareous aggregate or pozzolanic materials are present. A more useful technique is petrographic examination (to ASTM C1324) of specially prepared thin-section specimens using the polarizing microscope, which provides direct optical identification of the ingredients present (Figure 8). In addition, the mix proportions can be more accurately determined during the petrographic examination procedure by point counting or image analysis techniques.

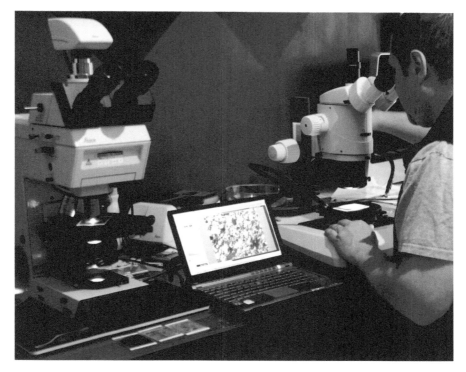

Figure 8   Petrographic examination of lime mortar using the low-power stereo microscope (right) and high-power petrological microscope (left). (Courtesy of IBIS Limited.)

The relative hydraulicity of hardened lime can be determined by microscopically observing the quantity of small relict grains of dicalcium silicate (also called 'belite'). Non-hydraulic limes have very few (if any) of these belite grains, with the number of grains observed increasing with hydraulicity. The most accurate determinations are yielded when reference samples are available for comparison. Petrographic examination will also determine whether any pozzolanic materials are contributing to hydraulicity.

DIAGNOSIS OF DEFECTS

The diagnostic process comprises analysis of all available information to form the most complete opinion as to the causes, extent and significance of the defects. This is an ongoing process that involves processing of data gathered at each stage of the investigation to narrow down the list of probable

Figure 9   Monitoring the growth of structural crack in Victorian brickwork by two different methods. A mortar pad is seen upper left and a plastic tell-tale is seen lower right, London, England.

defect causes in a systematic manner. A particular defect will often have more than one contributory cause, and the diagnostic process should aim to assess the relative significance of each cause to the overall failure.

Structural defects

Wide cracks that run through masonry units and render or plaster backgrounds are often structural in nature. If structural cracking is suspected it will be necessary to monitor cracks to determine whether they are still active. It used to be common practice to monitor the growth of cracks in masonry by applying pads of cement mortar or gypsum plaster to the surface of the crack at suitable locations, with the date inscribed on each pad (Figure 9). Further development of the crack fractured the pad and, providing the pad had not become detached, the resulting gap could be measured with reasonable accuracy. Today, either the Avongard tell-tale or the demountable mechanical strain gauge (DEMEC) are most commonly used for monitoring movement in cracks. The tell-tale consists of two plates of transparent plastic which are fixed across a crack so that they overlap for part of their length (Figure 9). Crack movements are measured by comparing the hairline cursor of the upper plate with the calibrated grid of the lower plate. The DEMEC technique involves using a demountable mechanical strain gauge to measure the distance between stainless steel discs affixed on either side of the crack.

Shrinkage

Cracks that are found to be confined to plaster or render layers, and are not present in the background, are often caused by shrinkage. They can often be found following a linear feature in the background such as the stud line in a timber lath background. Laboratory testing that shows high binder content, unsuitable aggregate or insufficient hair content also point to shrinkage as a cause of cracking.

The causes of surface crazing are often diagnosed by direct examination of samples in the laboratory, and can be indicated by excessive binder or a high fines content in the aggregate. Examination of the microscopical texture can indirectly diagnose over-trowelling, or inadequate curing.

Debonding and detachment

Diagnosing the causes of debonding or detachment of lime plaster and render typically requires scrutiny of site inspection records (including breakout records), and laboratory test data to eliminate the likely causes one by one. On-site breakout of the plaster or render will quickly establish if it was applied too thickly. Breakouts also allow examination of the background, and can indicate whether background movement, poor background preparation or poor suction control are involved.

The presence of weak material or secondary deposits may point to the involvement of leaching or a deleterious chemical reaction. Laboratory testing will establish if the plaster or render mix is too strong for the background, can check the quality of both ingredients and workmanship, and confirm if a deleterious reaction has occurred.

Low strength

Proof of the cause of low strength often comes from the findings of laboratory testing, such as observations of insufficient binder, inadequate mixing, poor compaction, uncarbonated binder or deleterious chemical reactions. Diagnosis can otherwise be inferred from a combination of site observations and laboratory test findings. For example, if a highly porous

Figure 10 Leaning brickwork chimney of a seventeenth-century mansion house, caused by sulfate action, Epsom, England.

background is found on site, and the laboratory test samples contain abundant desiccation microcracks, it is likely that suction control and curing conditions were inadequate. Diagnosing the causes of low strength of new construction can be considerably helped by knowing the site weather conditions during the building period. If construction took place at particularly hot or cold times of the year, it is likely that inadequate curing or frost damage may be clear contributory causes.

A need to diagnose the causes of weakening of lime mortars in historic structures is often brought about by concerns regarding the stability of structures. Figure 10 shows a brickwork chimney on a seventeenth-century manor house that was leaning sufficiently to cause concern over the potential for collapse. An investigation was conducted to determine the cause of the leaning. This involved visual inspection, removal of mortar samples, and limited laboratory testing. Inspection on site showed that the chimney was leaning away from the prevailing direction of driving rain, and that the original lime mortar had been repointed with inappropriate Portland cement mortar. This cement mortar had become extensively cracked, allowing moisture into the brickwork. Laboratory testing showed that the sulfate content of mortar samples from the exposed side of the chimney (4% by mass of binder) were twice that of samples taken from the sheltered side of the chimney (2% by mass of binder). It was concluded that sulfate contamination from the chimney flue had been involved in an expansive reaction causing swelling of the mortar joints on the most exposed side of the chimney, causing it to lean over towards the sheltered side. The failure in weatherproofing of the Portland cement repointing mortar was a major contributing factor to the occurrence of the sulfate action.

Fire damage

Fire damage is often apparent by its proximity to the seat of the fire, as revealed after the incident. As a rule of thumb, fire damage decreases in severity as you move away from the seat of a fire. Visual observations of spalled, cracked and/or discoloured material are commonly reported during investigations of fire-damaged masonry. Microscopical examination in the laboratory can confirm the temperature that the material has been subjected to, and the depth of resulting damage.

Staining and soiling

The causes of certain forms of surface staining and soiling such as pollution, organic growth, leaching and salt efflorescence may be readily apparent from visual inspection on-site. However there is often benefit in performing laboratory tests to confirm the site observations, as more specific information can aid the specification of suitable cleaning methods. For example, it is useful to know whether soiling is water-soluble. Laboratory testing of both the staining materials and the substrate is crucial to understanding the nature of any staining that does not have an obvious cause.

Diagnosing the causes of failure in coatings, such as paints, typically involves combining site inspection observations, laboratory test data and, if available, product data sheets for the coating product. Typical reasons for coating failure include the inappropriate use of an impermeable coating; inadequate substrate preparation; incorrect coating thickness or number of layers, or inadequate curing.

REPAIRING DEFECTS

Sources for guidance

Prior to repair, standard practice is to carry out investigations to identify the types of materials present, and the causes of decay. The purpose of these investigations is to ensure the compatibility of repair materials in terms of their physical, chemical and aesthetic characteristics, and to gain an understanding as to how future decay can best be prevented. Once the defects are adequately understood specialist advice should be sought from an architect, architectural conservator or engineer experienced in the repair of historic structures.

Useful guidance for the repair of masonry structures is provided in BS 8221-2.[9] BS EN 13914-1 and BS EN 13914-2 provide guidance for the restoration of renders and plasters on old and historic buildings.[10,11] Building limes should be specified in accordance with BS EN 459.[12] Additional guidance can be found in publications such as those by Ashurst, Holmes and Wingate, and Yates and Ferguson.[13,14,15]

Typical repairs

Table 3 summarizes the typical repairs required for a number of common defects in lime-based construction materials. The purpose of repair is typically to reinstate the structural capacity and weatherproofing of the structure to ensure its both safety and durability.

The type of repairs required for cracks depends on the causes, configuration and extent of the cracking. If structural cracks are found to be active it will be necessary to stabilize the structure before repairing any defects to lime-based construction materials. This will require the advice of a suitably experienced structural engineer. Stabilization may variously involve underpinning, grouting, stitching or the use of ground anchors. Inactive or stabilized cracks that run through the masonry units can be stitched back together by replacing cracked masonry units or installing dowel bars across the cracks.

For cracks that only run through mortar joints or through plaster and render layers it is usually sufficient to repoint the joints and/or cut out and replace plaster or render along the crack. Surface crazing patterns can be filled with lime slurry, unless unsoundness of the plaster/render makes replacement necessary.

Debonded or weak plaster and render is typically repaired by removing

Table 3 Typical repairs for selected defects in lime-based construction materials.

Defect Type	Repair
Structural cracking	Check if movement is still active. If stable, repair cracks and reinstate limework
Other cracking	Depending on extent, open out crack and fill with compatible mortar
Surface crazing	If plaster/render sound then fill hairline cracks with lime slurry. If unsound then remove and replace
Debonded/detached plaster or render	Remove loose material and replace. Consolidation grouting may be considered
Low strength, surface scaling/ friability	Partial or total repair with appropriate lime mortar/ plaster/render. Any underlying causes of deterioration (such as sulfate attack) should usually be addressed as part of the repair scheme
Fire damage	Remove and replace significantly weakened material. Alternatively, appropriate consolidation techniques may be considered
Coating failure	Remove failed coating and replace with appropriate coating, once any underlying causes of failure have been addressed
Staining	Remove staining/soiling using an appropriate cleaning technique. Consider addressing underlying causes of staining

the loose material back to the substrate, and then reinstating it with appropriate plaster or render mixes. Where there is an overriding requirement to conserve as much of the original fabric as possible it may be more appropriate to consolidate lime mortars, plasters or renders using techniques such as grout injection, resin impregnation, pinning or lime-watering. The use of such techniques normally requires the involvement of an experienced architectural conservator.

Repairs for weak or friable mortar joints will typically involve repointing with an appropriate mortar. In cases where there is severe deterioration it may be necessary to carefully dismantle the masonry and rebuild. However, due consideration should always be given to addressing any underlying reasons for the deterioration in order to prevent early re-occurrence of the same defects. For example, if sulfate attack has been caused by flue gases in a masonry chimney, then it would be prudent to remove sulfates deposited in the masonry by replacing contaminated mortar, and possibly poulticing the bricks, and to prevent future sulfate contamination by installing a suitable chimney liner.

Lime that has been significantly weakened by heating in a fire would normally be removed or consolidated. This includes reddened materials, as this normally indicates that these are significantly weakened. Surfaces with black soot deposits are typically cleaned using an appropriate technique that will not damage the substrate. This applies to other staining materials such as pollution crusts, algae/lichen, bird fouling, lime weeping and rust. It is advisable to remove impermeable paint coatings, and if appropriate to replace them with compatible, permeable coatings. A good starting point for advice regarding suitable cleaning techniques for removal of staining and soiling, and for removing inappropriate paint coatings and graffiti, is BS 8221-1.[16] Again, however, the advice of a specialist architectural conservator is likely to be needed to ensure that cleaning techniques do not damage valuable remaining fabric.

Choice of ingredients

The choice of ingredients and mix proportions for repair mortars requires careful consideration. Factors to be considered include matching the physical, chemical and aesthetic properties of the existing material, the exposure conditions, and the state of conservation of the structure. Compatible repair materials may be chosen to provide an exact or close match to pre-existing material, or to provide increased local durability without compromising the durability and aesthetics of the structure as a whole. To aid the process of locating aggregate and lime sources in England, lists of suppliers and test data for some currently available building sands and lime products suitable for use in restoration have been published.[17,18]

Selection of contractor

To ensure successful repair it is advisable to use an experienced contractor, and to require that they provide full and detailed method statements. It is prudent to request submittal of materials samples with manufacturers' data sheets, to undertake trial works, and ensure that slow curing of lime mortars is undertaken. As with the initial investigations, appropriate consents and permissions must be obtained prior to undertaking any repairs. When devising repair schedules, due consideration should be given to providing repair solutions that can be implemented safely. In the United Kingdom, construction activities must be conducted in accordance with the Construction (Design and Management) Regulations 2007.[19] There are a number of health and safety issues involved in working with lime. In particular, quicklime is extremely caustic and fresh lime mortar only slightly less so. Also, the slaking process generates heat, causing steam and splashing. At all stages lime products require careful handling and operatives should make appropriate use of personal protective equipment. In 2007 the Registration, Evaluation, Authorisation and Restriction of Chemicals (REACH) regulations came into effect throughout the European Union. This requires companies making and supplying mortars and other lime products from quicklime to register with REACH.[20]

CONCLUSIONS

The lime-based construction materials that are present in many of our historic structures can suffer from a variety of defects, caused either by natural processes or by man. Understanding the causes and extent of these defects is an essential prerequisite to correct specification of repairs for historic structures. This can be successfully achieved by a combination of desk study, on-site condition surveys and laboratory testing of samples. The correct application of these techniques, combined with careful diagnosis, will undoubtedly provide a valuable contribution to the conservation of our built heritage.

This paper was first published in the *Journal of Architectural Conservation*, Volume 15, No. 3, November 2009. This is a revised and updated version.

Notes

1   McKibbins, L. D., Melbourne, C., Sawar, N., and Sicilia Gaillard, C., 'Masonry arch bridges: condition appraisal and remedial treatment', *CIRIA C656*, CIRIA, London, 2006.
2   Ingham, J. P., 'Laboratory investigation of lime mortars, plasters and renders', *Journal of The Building Limes Forum*, Vol. 10, 2003, pp. 17–36. Reprinted in this volume.
3   Ingham, J. P., 'Investigation of traditional lime mortars – the role of optical microscopy', *Proceedings of the 10th Euroseminar on Microscopy Applied to Building Materials*, Paisley, June 2005, pp. 21–5.
4   Ingham, J. P., 'The role of light microscopy in the investigation of historic

masonry structures', *Proceedings of the Royal Microscopical Society*, Vol. 40/1, 2005, pp. 14–22.

5   Ingham, J. P., 'Forensic engineering of fire-damaged structures', *Proceedings of the Institution of Civil Engineers, Civil Engineering*, 162, Special issue – Forensic engineering, May 2009, pp. 12–17.

6   Ingham, J. P., *op. cit.*, 2003.

7   British Standards Institution (BSI), *BSI 4551. Mortar – Methods of test for mortar. Chemical analysis and physical testing*, BSI, London, 2005.

8   ASTM International, 'Standard test method for examination and analysis of hardened masonry mortar', ASTM C1324-10, Philadelphia, USA, 2010.

9   British Standards Institution (BSI), *BS 8221-2. Code of practice for cleaning and surface repair of buildings – Part 2: Surface repair of natural stones, brick and terracotta*, BSI, London, 2000.

10 British Standards Institution (BSI), *BS EN 13914-1. Design, preparation and application of external rendering and internal plastering – Part 1: External rendering*, BSI, London, 2005.

11 British Standards Institution (BSI), *BS EN 13914-2. Design, preparation and application of external rendering and internal plastering – Part 2: Design considerations and essential principles for internal plastering*, BSI, London, 2005.

12 British Standards Institution (BSI), *BS EN 459-1. Building lime – Part 1: Definitions, specifications and conformity criteria*, BSI, London, 2010.

13 Ashurst, J., *Mortars, Plasters and Renders in Conservation (2nd edition)*, Ecclesiastical Architects and Surveyors Association, London, 2002.

14 Holmes, S. and Wingate, M., *Building with Lime: A Practical Introduction (2nd edition)*, ITDG Publishing, 2002.

15 Yates, T. and Ferguson, A., *NHBC Research Paper NF12: The use of lime-based mortars in new build*, NHBC Foundation, Milton Keynes, 2008.

16 British Standards Institution (BSI), *BS EN 8221-1. Code of practice for cleaning and surface repair of buildings – Part 1: Cleaning of natural stones, brick, terracotta and concrete*, BSI, London, 2000.

17 English Heritage, *The English Heritage Directory of Building Sands and Aggregates*, Donhead Publishing Ltd, Shaftesbury, 2000.

18 English Heritage, *The English Heritage Directory of Building Limes*, Donhead Publishing Ltd, Shaftesbury, 1997.

19 The Construction (Design and Management) Regulations 2007 (Statutory Instruments), Stationery Office Books, 2007.

20 Regulation (EC) No 1907/2006 of the European Parliament and of the Council of 18 December 2006 concerning the Registration, Evaluation, Authorisation and Restriction of Chemicals (REACH), establishing a European Chemicals Agency.

To Wake a Gentle Giant

Grey Chalk Limes Test the Standards

Stafford Holmes

VANISHING STANDARDS

Methods of lime burning and production today, with technically advanced kilns, result in building limes that differ from those used before the twentieth century. The source of limestone, kiln design, and the way mixes are prepared and tended all have a significant effect on the end result.

Fortunately, archive records and technical references are plentiful. They confirm the changes that have taken place in building practice and the use of building limes over the centuries. The most alarming of these changes occurred in 2002, when grey chalk lime and other slightly hydraulic limes were omitted from the British Standards due to the withdrawal of BS 890.

References to the wide range of building limes, including grey chalk lime, and their chemical composition and appropriate application have been given in numerous textbooks until quite recently, and historically by Smeaton, Vicat and Cowper.[1,2,3] It is clear from these and many other records that grey chalk lime and other slightly hydraulic limes are an important part of the wide spectrum of building limes traditionally produced and extensively used in Britain. The various types of building lime range from pure, fat, 'air', non-hydraulic limes to eminently hydraulic limes and natural cements. The majority of limes in general use until the last half of the twentieth century were well below the strengths of those given in the current Standards. New imported hydraulic limes, which are now rapidly gaining in popularity, can result in stronger mortars than the 1:1:6 and 1:2:9 mixes containing Portland cement.

The significance and widespread application of grey chalk and slightly hydraulic limes is confirmed by:

BS 890: 1972

This Standard, now withdrawn, sets out requirements for lime made from limestone and chalk under sub-paragraph (b) for 'semi-hydraulic lime (grey lime)'. References to grey chalk lime also occur on pages 9 and 10 of this Standard.[4]

Mitchell's Materials 1970–94

This textbook states that, 'Semi-hydraulic lime is derived mainly from the Grey Chalk of the southern counties of England hence the term Grey Stone Lime.'[5]

It is curious that grey stone lime is derived from chalk and not stone. One of the reasons for this may well be that some of the lower chalk beds are hard enough to be used in the form of ashlar masonry, as is the case with clunch. Many examples of building with clunch, including the external walling of historically important country houses, remain in existence today.

Grey chalk lime manufacture

As recently as the mid-twentieth century the Southern Lime Association had at least 20 manufacturing members who produced grey chalk lime. These included Dorking Greystone Lime Co. Ltd, Merstham Lime Co., Oxted Greystone Lime Co. Ltd and Totternhoe Lime and Stone Co. Ltd.[6]

There can be little doubt, therefore, that grey chalk and slightly hydraulic limes were recommended and extensively used for construction in Britain until the twentieth century. The production of building limes from chalk was normal practice and is documented from at least the thirteenth century.

HISTORICAL REFERENCES

There are sufficient records of provenance to determine not only the extensive use but also the quality, location and long history of this gentle giant of British lime. The following information is derived from many sources and I acknowledge with thanks the archive research by Mr L. F. Salzman, whose detailed recording has provided the medieval references.[7]

Building construction records in areas where the geological formation is chalk, particularly where masons are contracted to dig and burn their own lime, are relevant, as they reveal the extensive use of burnt chalk from various bed depths, and local lime for mortar.

A common theme running through medieval building contracts is that the mason is held responsible for finding the lime, the sand and the stone, and for building his own lime kiln and burning the lime. A specification for the mortar to be used at Eton College Chapel in around 1453 leaves no doubt about the quality required. It states, 'Good and myghty mortar is to be made with fyne stone lyme and gravell sonde'.

In his letter dated 7 June 1549, the clerk of works to Sir John Thynne advises on the selection of stone for a house being built at Bedwyn, Wiltshire, and the quality of the stone from Wilton Quarry. It appears that different chalk beds were accessible. The description by the clerk of works is revealing. The original wording is given here but current spelling has been used for clarification. The importance of this record is that it confirms that the different qualities of the various chalk beds were well understood. The harder, lower chalk beds were selected for ashlar, while the upper beds and broken pieces were burnt for lime mortar.[8]

> The stone of Wilton Quarry makes very good lime ...; and whereas there is a great heap of dust made by means of the rubble which came out of the quarry, the same will serve very well to mingle with the lime in the filling of the walls; for it is of itself very tough when it is beaten and tempered and much more it will be tough when the lime helps it to bind. Thus does the best of the stone make good ashlar, and the ragstone will serve for the foundations and infilling stuff, and the rubble for lime and the dust for rough work, so that whatsoever cost my Lord's grace bestows there, it will quit the cost.

Sir William Petty wrote to the Philosophical Society of Oxford in 1684, confirming that the lime used in making various types of mortar was generally of two sorts: 'chalk-lime', made from chalkstone dug at Nettlebed, and other places, and burnt; and, second, hard stone lime, made from burnt hard ragstone, which is much stronger.[9]

John Smeaton

By 1759 the Eddystone lighthouse was completed, and although Smeaton's narrative of the building and construction of the lighthouse was not

Table 1   Limestones analysed by Smeaton and tabulated on page 117 of his book on the design and construction of the Eddystone lighthouse. [Percentage conversions of the proportion of clay, originally given in fractions, added by the author.]

No.	Species of Limestone	Proportion of Clay	Percentage of Clay	Colour of the Clay
1	Aberthaw	$^3/_{23}$	13.0	Lead
2	Watchet	$^3/_{25}$	12.0	Lead
3	Barrow	$^3/_{14}$	21.4	Lead
4	Long Bennington	$^3/_{22}$	13.6	Lead
5	Sussex Clunch	$^3/_{16}$	18.7	Ash
6	Dorking	$^1/_{17}$	5.9	Ash
7	Berryton Grey Lime	$^1/_{12}$	8.3	Ash
8	Guildford	$^2/_{19}$	10.5	Ash
9	Sutton	$^3/_{16}$	18.7	Brown

published until 1793, his research into building limes was carried out in 1757. Smeaton refers to lime from Dorking ('Darking') in Surrey, which is brought to London. He states that those purchasing the Dorking lime are under the impression that it is burnt from stone, and this is the reason that it is stronger than the chalk lime in common use in London. Smeaton states that in fact Dorking lime is made from chalk which is not much harder than common chalk. He analysed the Dorking grey chalk, and the results from the sample he selected showed that it contained $^1/_{17}$ part (5.9%) of light-coloured clay.[10] This supported his contention that the hydraulic qualities of a lime are due to its clay content.

As part of Smeaton's research to develop appropriate mortars for the lighthouse, he examined a wide range of building limes regularly used in Britain at that time, including grey chalk limes. The narrative describes his research on the grey chalk lime from Berryton, near Petersfield in Hampshire, which was transported to Portsmouth for use in hydraulic lime mortar. He states that the Berryton lime is close in composition to Sussex clunch and that it appears similar to Dorking lime, although it has a greater clay content.

The narrative refers to different chalk pits in the neighbourhood of Guildford, and particularly the grey chalk used for Guildford Lock. The mortar joints to the lock walls were described as being in sound condition, especially where they were frequently under water. Smeaton's analysis of the grey chalk in this area shows that it contained $^2/_{19}$ part (10.5%) of dark clay. He goes on to propose that this is useful information for London builders. Smeaton develops his thesis on grey chalk limes by suggesting, quite correctly, that similar grey stone lime may be found along the range of chalk hills from Lewes to Petersfield, 'and probably from thence to Surrey, to Guildford and Dorking'.[11] Smeaton set out a table of his analysis which shows the clay content of the grey chalk limestones from Dorking, Berryton, Guildford and Sutton.[12]

THE NINETEENTH CENTURY

Geological mapping

The investigation and research of natural resources, particularly rocks and minerals, gathered pace in the nineteenth century due to the enquiring minds of dedicated laymen, scientists and engineers. The location of these beneficial and often valuable materials, and the way they had been formed, were little known or understood.

The geological mapping of Britain by William Smith and the production of the first geological map in 1815 was a major breakthrough. William Smith had single-mindedly and single-handedly investigated the geology of Britain, although his achievements were not fully recognized until 1829. Geological maps to this day are an extensive source of information, which has been of benefit worldwide. They may also be used retrospectively for the subsequent confirmation of earlier opinions held by engineers such as Smeaton, in connection with the location of grey chalk lime.

Mapping of the Cretaceous beds assisted those wishing to locate and exploit chalk, as the demand for grey chalk lime in London steadily increased. F. J. North states that by 1820 most of the lime used in London was made from chalk from the neighbourhood of Gravesend.[13]

Grey chalk lime used in London

In 1854, John Weale publishes a series of technical books on the art of building. These confirm that the sources of building lime used in London are those obtained from the beds of the lower chalk. Dorking and Halling limes near Rochester, Kent, are the principal limes used in London for making mortar, and are slightly hydraulic. It is also confirmed that the mortar made with these limes sets hard and when set may be exposed to considerable moisture without damage. The limes will not, however, set under water unless combined with a pozzolan.[14]

In 1867 George Burnell states that, 'the very moderately hydraulic limes used in London' are adequate for buildings above ground, for which the common Dorking or Halling limes are used. He also confirms that the normal method of preparing mortar is by throwing water over the lime and then covering it with sand.[15]

The New Guide to Masonry, Bricklaying and Plastering by Robert Scott Burn describes the results of Smeaton's experiments on English hydraulic limes, and refers to grey chalk limes from Dorking, Berryton and Guildford, as well as Sussex clunch from near Lewes.[16] *Gwilt's Encyclopaedia of Architecture*, published in 1851 and revised and reprinted in 1894, states that among the strongest limes that will set under water and are mostly used in the Metropolis, are those called grey stone limes, procured from Dorking, Merstham and the vicinity of Guildford in Surrey. The Dorking and other limes in that area are burnt from a chalk formation which is extremely hard, and is hard enough to be quarried for the purpose of masonry. Those of Merstham, particularly, are from an 'indurated' (hardened) chalk marl (clay and chalk) 'which is so hard that it partakes of the nature of stone'.[17]

It would appear from these various descriptions that grey chalk lime may or may not set under water, and that its setting is dependent upon the precise proportion of 'active clay' content in the particular bed selected. It also appears that, once set, grey chalk lime mortar has considerable durability in wet conditions.

Rivington's series of *Notes on Building Construction* states under 'Hydraulic Limes':[18]

> Grey chalk lime (called 'stone lime' in London) is of feebly hydraulic character. It is obtained from the lower chalk beds in the South of England, the present supplies coming from Halling, Dorking, Lewes, Petersfield, Merstham, etc. This lime is usually of a light buff colour, and slakes very freely. When used with two parts of sand in brickwork, a good sample should sensibly resist the finger-nail at a month old.

A chemical analysis of the Halling grey chalk attributed to Col Scott shows that the sample tested had an 8% clay content.[19]

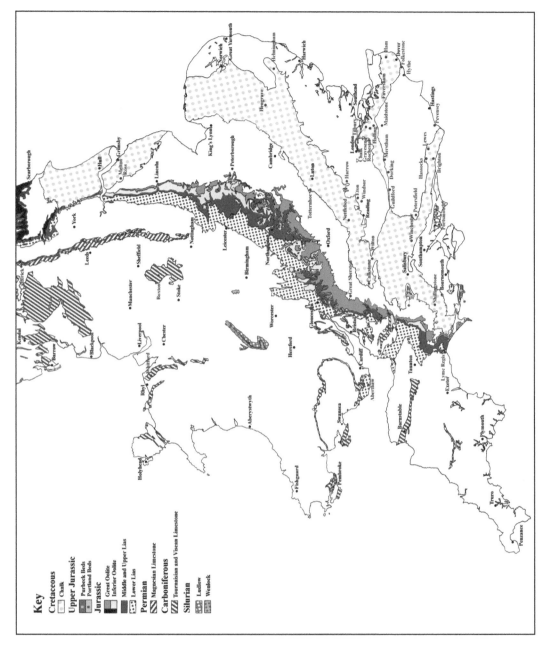

Figure 1
Principal limestone formations in England and Wales south of Catterick. Based on information from the IGS Geological Survey.

THE TWENTIETH CENTURY

Textbook references and specifications

In 1908 issue 11 of *Specification*, incorporating the Municipal Engineer's Specification for Architects, Surveyors and Engineers, included grey stone lime in the materials section. It is described as slightly hydraulic. The description also states that it is obtained chiefly at Dorking, Halling, Lewes, Merstham etc.[20] In the section on mortar it is described as '*stone lime* which is feebly hydraulic, will make a fair mortar, but should not be generally used for work below ground'.[21] In the sample specification for lime mortar it states:[22]

> The stone lime mortar for brickwork above ground level shall be composed of one part *grey chalk lime* ... and two to three parts of sand, mixed with a sufficiency of water and thoroughly incorporated together (in a mortar mill). (The lime and sand shall be mixed together in their dry state before being put into the mortar mill.)

Specification was updated every year until publication ceased in the 1990s, and changes in terminology can be followed. Both white chalk lime and grey chalk lime were used in London. The grey chalk produced a slightly hydraulic lime which was used extensively for external mortars. The white chalk lime was used for internal plasterwork. The lower chalk beds produced the grey chalk, some of which, such as clunch, was hard enough to be used as ashlar. This again indicates the reason for the general belief in London that the hardest stones would produce the most hydraulic limes. The terms 'stone lime' and 'grey stone lime' became synonymous with 'grey chalk lime'.

Standardization

In an attempt to create a standardized description, the slightly hydraulic characteristic of grey chalk limes was attributed to other building limes with similar levels of hydraulicity, and the term 'semi-hydraulic lime' was coined. This can at best be described as ambiguous!

Thus the terms *grey chalk lime, grey stone lime, greystone lime, stone lime, lean lime, semi-hydraulic lime, slightly hydraulic lime* and *feebly hydraulic lime* have all become interchangeable. No wonder there has been confusion! Local terminology and regional dialects probably add to this list.

APPENDIX

THE SOUTHERN LIME ASSOCIATION

LIST OF MANUFACTURERS OF LIME
FROM WHITE AND GREY CHALK

Manufacturer	Situation of Works and Telephone No.		Map Ref.
ANGLO-ROMAN LIME LTD.	Watling Street, St. Albans, Herts.	Redbourn 238	
	Swann Works, Cherry Hinton, Nr. Cambridge.	Cambridge 87987	1
BARNET LIME CO. LTD.	Barnet By-Pass, South Mimms, Herts.	South Mimms 2154	2
BARTON (BEDS) LIME CO. LTD. 91, Cadogan Lane, London, S.W.1. (Sloane 8203)	Barton-in-the-Clay, Beds.	Hexton 215	3
BRIGHTON ROCKLIME WORKS	Poynings, Nr. Brighton, Sussex.	Poynings 59	15
CEMENT MARKETING CO. LTD. 192, Ashley Gardens, London, S.W.1. (Victoria 6677)	Dunstable Works, Dunstable, Beds. Holborough Works, Snodland, Kent. Johnsons Works, Dartford, Kent. Wouldham Works, Grays, Essex.	Dunstable 194 Snodland 84266 Dartford 3005 Tilbury 10	4
DANE JOHN LIME-WORKS LTD.	2, Watling Street, Canterbury, Kent.	Canterbury 4275	5
DORKING GREY-STONE LIME CO. LTD.	Betchworth, Surrey.	Betchworth 3317	6
DUNSTABLE VALLEY LIMEWORKS.	Lime Kiln Lane, Kensworth, Nr. Dunstable, Beds.	Whipsnade 78 & 52	7
HADHAM LIME WORKS LTD.	Lime Kiln Lane, Albury Road, Little Hadham, Herts.	Albury 260	8
HALL & CO. LTD., Victoria Wharf, Croydon, Surrey. (Croydon 4444)	Marlpit Lane, Coulsdon, Surrey.	Uplands 5157	9
HITCHIN LIME WORKS LTD.	Cadwell Lane, Hitchin, Herts.	Hitchin 598	10
IVINGHOE LIME CO. LTD.	Ivinghoe Aston, Leighton Buzzard, Beds.	Eaton Bray 211	11
MERSTHAM LIME CO. LTD.	Merstham, Surrey.	Merstham 9	12
NEWINGTON G. & CO. LTD. 14a, Eastgate Street, Lewes, Sussex.	Glynde, Sussex.	Lewes 5	13
NORTH DOWNS LIME CO. LTD.	Hogtrough Hill, Brasted, Nr. Seven-oaks, Kent.	Brasted 185	13a
OXTED GREYSTONE LIME CO. LTD.	Oxted, Surrey.	Oxted 31	14
STORTFORD LIME WORKS LTD.	Lime Kiln Lane, Farnham Road, Bishop's Stortford, Herts.	Albury 219	16
TOTTERNHOE LIME & STONE CO. LTD.	Totternhoe, Nr. Dunstable, Beds.	Eaton Bray 300	17
BUTSER HILL LIMEWORKS LTD.	Portsmouth Road, Petersfield, Hants.	Petersfield 790	18
PAULSGROVE LIME WORKS.	Paulsgrove, Nr. Cosham, Portsmouth.	Cosham 76012	19

Figure 2 Page from a publication by the Southern Lime Association, believed to date from *c.* 1950, listing members who produced grey chalk lime at least until the mid-twenteith century.

Research

A report on the composition and strength of mortars was published in 1911. This is a comprehensive report on the results of the experimental investigation conducted by W. J. Dibdin for the Science Standing Committee of the Royal Institute of British Architects. Experiments were carried out over two years to determine the relative strengths of mortars composed of building limes and sands at varying mix proportions ranging from 1 lime:2 sand to 1 lime:5 sand.

Dibdin's experiments included mortars made with Dorking grey stone lime. The analysis of Dorking grey stone lime in his report shows this as having 5.79% silica and 3.62% iron and alumina.[23] The compressive strengths of mortar mixes using one volume of Dorking grey stone lime to two volumes of standard (well-graded) sand ranged from 153 lb per square inch at one month to 257 lb per square inch at two years (1.04 N/mm² to 1.75 N/mm²). One year before this, John Howe published an analysis of grey chalk lime from Folkestone, the result of which shows that it included 3.61% silica and 1.29% alumina and iron.[24]

Classification

In 1927 A. D. Cowper completed the BRS Special Report No. 9, entitled *Lime and Lime Mortars*, for the Building Research Station.[25] This is sufficiently important to have been reprinted by Donhead in 1998. In the foreword to the 1998 reprint, Tim Yates of the BRE describes the report as a comprehensive and state-of-the-art review which remained a valuable source of information and guidance for many years. Cowper adopts and develops Vicat's method of classifying building limes and includes this as a table in the report. He modifies the classification slightly and proposes a number of classes, including Class B for poor or lean limes and Class C1 for slightly or feebly hydraulic limes.

Cowper confirms that the Cretaceous deposits in Yorkshire and the south-west contain an upper layer of white chalk, producing a Class A lime. Below this is frequently a layer of grey chalk, containing some argillaceous (clayey) matter, and yielding a lime corresponding to either Class B or Class C1. Below this again occurs the chalk marl, 'clunch'. There is a footnote confirming that clunch is a generic term applied to all chalk rocks that can be used for building purposes. This seems to be further confirmation that the term 'stone lime' developed due to the association with the more hydraulic and stronger chalks, some of which were used for building stone.

RECENT REFERENCES AND STANDARDS

Limestone and its Products, 1935

The author of *Limestone and its Products*, A. B. Searle, categorizes the Cretaceous system in the UK into three main groups. These are: the upper chalk, generally white with flints; the lower or grey chalks, usually containing 10% or more clay which makes them slightly hydraulic; and chalk marl, which is a greyish or yellow mix of chalk and clay. Searle confirms that the lower or grey chalk is mostly used for building lime and is known locally as grey stone lime or flair lime. The term *flair lime* refers to the type of kiln in which it is burnt. The Dorking, Halling and Medway limes are from this formation. Chalk marl was sometimes used for making clunch lime, although this was of uncertain quality.[26]

Searle also states that grey stone limes are highly appreciated for the strength of the mortar which they produce. One month is required after ground grey stone lime is mixed with water before it develops appreciable hydraulic strength. It loses this property if it is stored in the form of putty or if it is soaked or kept for a long time before use. Some of the locations at which the lower chalk beds were worked are mentioned, and these include Maidstone, Totternhoe, Hythe, Hassocks and Calkstone.[27]

McKay's *Building Construction*, 1938–1952

W. B. McKay defines and describes grey chalk lime as a lean, poor, grey chalk or stone lime containing more than 5% of clay impurities. There follows a typical analysis that shows the combined total of silica, alumina, and iron oxide as over 11%.[28] McKay's *Building Construction* was first published in 1938, the second edition in 1943, and the third in 1952. It was recently reprinted, in 2005.

BS 890: 1940–1972

In 1940 BS 890 was published. This included Class A limes for plastering, finishing coats and building mortar, and Class B for coarse stuff and building mortar. The Standard confirms that either class may be of the non-hydraulic or semi-hydraulic types. The compressive strength of semi-hydraulic lime in a 1 : 3 mortar by weight at 28 days' curing at 25°C is given as 100 lb per square inch minimum and 300 lb per square inch maximum. This equates to 0.68 N/mm² minimum and 2 N/mm² maximum.[29]

BS 890 was reprinted in 1966. This edition also states on page 3 the requirements for lime made from limestone and chalk. There are three categories of building lime that are dealt with in the Standard and these are described as:[30]

1. high calcium lime (white lime);
2. semi-hydraulic (grey lime);
3. magnesium lime.

The metric edition of BS 890 was published in October 1972. This edition again retains the description 'semi-hydraulic lime (grey lime)' as well as the minimum and maximum compressive strengths converted to the metric equivalents of 0.7 N/mm² and 2.0 N/mm².[31]

Mitchell's Building Construction, 1978–1986

In 1978 Alan Everett states in *Mitchell's Building Construction* that semi-hydraulic lime is derived mainly from the grey chalk of the southern counties of England, hence the term 'greystone lime'.[32] *Mitchell's Materials* 5th edition, revised by C. M. H. Barritt and published in 1986, repeats the definition of semi-hydraulic lime as a lime derived from the grey chalk of southern England and termed greystone lime.[33]

DEFINING GREY CHALK

Attempting to give a precise definition of a natural material has its risks. It is a challenge in most cases and not least in the case of grey chalk. There is sufficient information, however, to be able to propose some parameters for grey chalk in terms of geology, chemistry, setting properties and classification.

Geology

William Smith chose blue-green for the chalk when colouring his original geological map in 1815.[34] This is a colour not far from the green used for chalk on the Geological Survey maps today. Thanks to the recognition and development of William Smith's work, therefore, the location of the main chalk formations is not in doubt. The chalk beds are composed of numerous strata which extend from the south of England to Yorkshire. The principal beds in England are simply described as upper, middle and lower chalk. It is from the lower chalk beds that the grey chalk limes are obtained. Chalk occupies a larger area than any other limestone formation in the British Isles, excepting the carboniferous limestone series. In 1930 close to one-third of all calcareous rock quarried in the UK was chalk. Home Office records at the time listed over 200 chalk quarries, with nearly half of the total UK output from Kent.[35]

Chemistry

F. J. North provides a chemical analysis of grey chalk from Folkestone which gives this as 0.31% magnesium carbonate, 0.29% iron and alumina, 3.61% silica and insoluble residues, and 94.09% calcium carbonate. An average analysis of Kentish grey chalk published in 1935 gives the following:

Silica	4.59%
Alumina	2.64%
Iron	1.33%
Magnesium oxide	0.29%
Calcium oxide	50.7%
Loss on ignition	40.45% [36]

Mr R. A. Bates of The Totternhoe Lime & Stone Company Limited has kindly provided a representative analysis of their grey chalk, which was burnt until January 1993. This is:

Total volatile matter	1.9%
Silica	8.7%
Iron oxide	1.3%
Aluminum oxide	3.5%
Lime ($CaCO_3$)	82.8%
Magnesia	0.9%
Sulphuric anhydride	0.4%

In addition to the above, there are the analyses provided by Smeaton, Vicat and Cowper. The term 'active clay content' was coined by Vicat as a simple way of describing the active silica, alumina, iron and other minerals as a total percentage of the limestone which acts as the hydraulic component of the lime after burning.[37]

Setting properties

The way grey chalk lime sets in a mortar can be verified by the craftsmen that have used it. Fortunately there are also written records of this. Searle advises that grey stone limes are highly appreciated for the strength of the mortar which they produce as a result of being feebly hydraulic. About one month is required after they have been mixed with water before they develop appreciable hydraulic strength. Rivington states that when the lime is used with two parts of sand in brickwork it should resist the fingernail at a month old. Cowper refers to Vicat's table, which indicates that the consistency of Class C1 limes after one year in water should be intermediate between putty and hard soap and that the set (out of water) should be within 20 days.

Classification

The classification for grey chalk lime recommended by Cowper is C1. This implies an active clay content of below 12%, or not more than 15% silica and alumina with not less than 60% calcium and magnesium oxide. BS 890 requires the compressive strength of semi-hydraulic grey lime mortar to be from 100 lb per square inch minimum to 300 lb per square inch maximum. This is converted in the revised (but now withdrawn) metric edition of BS 890, which gives from 0.7 N/mm^2 minimum to 2 N/mm^2 maximum.

The missing link

In 2002 BS 890 was withdrawn, to be replaced by BS EN 459, 'Building Limes'. All references to grey chalk and slightly hydraulic limes are omitted from the new Standard. References to compressive strength test results between 0.7 and 2.0 N/mm^2 are also omitted. Hydraulic lime categories are introduced with broad, overlapping compressive strength requirements for classification, the lowest being NHL2 with a compressive strength requirement of between 2.0 N/mm^2 and 7.0 N/mm^2. The direct comparison in lb per square inch is minimum 300 lb per square inch and maximum 1,030 lb per square inch. The other classes of hydraulic lime in this Standard are NHL3.5 and NHL5. A summary of the effect of this change in the Standards for the compressive strength of lime mortar, using BS standard test methods, is set out in Table 2 (lbs per square inch abbreviated to psi).

Table 2 Comparison of fat and feebly hydraulic limes with NHL graded hydraulic limes to BS EN 459:2002. All mortars for these tests are mixed in the ratio 1 lime to 3 sand by mass, in accordance with the Standard.

Traditional description	Current description	Compressive-strength requirement (psi)	Compressive-strength requirement (N/mm²)	Status
Fat lime Lime putty	Non-hydraulic 'pure', 'fat', 'air' limes	None	None	BS EN 459 current
Feebly hydraulic	Grey chalk, semi-hydraulic (slightly hydraulic) limes	100 psi min. 300 psi max.	0.7 N/mm² min. 2.0 N/mm² max.	BS 890 withdrawn in 2002. The missing link.
Moderately hydraulic	NHL2	300 psi min. 1,030 psi max.	2.0 N/mm² min. 7.0 N/mm² max.	BS EN 459 current from 2002
Eminently hydraulic	NHL3.5	515 psi min. 1,427 psi max.	3.5 N/mm² min. 10.0 N/mm² max.	BS EN 459 current from 2002
Natural cement	NHL5	736 psi min. 2,208 psi max.	5.0 N/mm² min. 15.0 N/mm² max.	BS EN 459 current from 2002
Compare	Portland cement: lime: sand mortar to BS 5838	441.6 psi min.	3.0 N/mm² min.	BS 5383

The Portland cement:lime:sand mortar prepared in accordance with BS 5838, given for comparison at the bottom of the table, is for pre-packed dry mixed bagged mortar to which only water has to be added. The Portland cement, lime and sand proportions are not stated in the Standard. The reason for including this in the table is to demonstrate that current standards incorporating lime and OPC binders require minimum strengths which are in excess of that previously stipulated in BS 890 for British hydraulic (grey chalk) lime; and yet grey chalk lime has been used successfully in external walling for centuries. Manufacturers of hydraulic lime are required to apply 'conformance criteria' when testing their products, which effectively limits the compressive strength compliance range for NHL2 to between 3.5 and 5.8 N/mm².

If existing, and often historically important, mortars are to be matched this cannot be achieved by preparing a mortar with a hydraulic lime or non-hydraulic and OPC mix in accordance with test results for the current standards. The best means of arriving at a compatible mortar is with a grey chalk lime which now has no standard, and is no longer commercially produced, although there are numerous quarries and vast deposits of it in this country.

The table also indicates the way in which increasing strengths of the present mass-produced hydraulic limes, particularly NHL5, are closer to the strengths of earlier OPC mortar mixes. The risk that the use of some of these new hydraulic limes may result in mortars stronger than the adjacent masonry units (brick or stone) is evident.

In addition, in respect of compressive-strength requirements, it should be noted that the current British Standard tests are carried out at 28 days. After the 28-day test the strength gain for lime mortar over one to two years has been shown to be as much as 90% or more, depending on mix ratios and free lime content. This is according to Dibdin and the recent Foresight Lime Research at Bristol University. Although it diminishes, there is a continuous strength gain for well over two years. This level of increase over a long period does not occur with the Portland cement mixes, as the majority of the strength gain (approximately 80%) takes place before the 28-day test period.

CONCLUSION

Hydraulic quality

The Cementation Index (CI), which determines the hydraulic quality of building limes, can be used to check all the grey chalk and slightly hydraulic limes for which chemical analysis exists. Details of the CI and the way it is applied can be found in *Building with Lime*, where the CI for slightly hydraulic limes is given as 0.3 to 0.5.[38] Chemical analysis of the majority of limestones used in the past can be traced back to the mid-eighteenth century if not earlier, thanks to Smeaton's thorough investigation. While some of the historical analysis gives the clay content only, which does not always differentiate between silica, alumina and iron, there is generally sufficient information to determine the degree of hydraulicity or approximate CI. Chemical analysis alone, therefore, is sufficient to confirm the extensive use to which grey chalk limes were put, particularly in and around London.

Appropriate selection and use

In London the fat and purer limes obtained from the upper white chalks and other pure high-calcium stone limes were preferred for internal work and plastering, while the grey chalks from the lower beds were preferred for external work. However, civil engineers designing structures such as bridges, docks, canals and lighthouses tended to be the exception to the rule, and they created an increasing demand for moderately and eminently hydraulic limes. For these, blue lias plus pozzolan was often specified in the nineteenth century.

All evidence points to the fact that the grey chalks generally were slightly hydraulic and ranged in strength between 0.5 and 2.0 N/mm² compressive strength tested in accordance with the British Standard test methods. The British Standard for the grey chalk lime compressive strength requirement was in force for over 60 years, until 2002. There has been comment that this test method was not satisfactory. If this is the case a different method of classifying grey chalks should be used, or the compressive test method should be changed.

There is evidence that slightly hydraulic limes were produced from limestone formations other than the Cretaceous beds. Grey chalk and slightly hydraulic limes were regularly used in masonry mortar and are appropriate for most buildings of up to three stories or more, in locations that are not subject to constantly wet conditions and severe exposure.

Revising the Standards

The importance of using appropriate materials in carrying out repairs to historic buildings and traditional construction is well understood. This is a fundamental requirement by specifiers, who now have no current Standard on building limes adequate for the purpose. The British Standards have, for many years, acted as a bedrock for specification writing. Professionals and contractors in the building industry have used them as a basis to agree the minimum quality of materials and workmanship expected.

If new Standards are inadequate and do not provide essential information for materials that must be, and are, specified, they will lose credibility. Problems have arisen due to the specification of 'hybrid' limes (a mixture of non-hydraulic and hydraulic lime) in the absence of a slightly hydraulic lime classification. Slightly hydraulic limes should therefore be reintroduced to the current British Standards, or, as an interim measure, BS 890 should be reinstated. This is of serious concern to specifiers and users of traditional building limes and is the case not only throughout Britain but probably also in many other parts of Europe.

Similar chalk formations and slightly hydraulic limes occur in France and other European countries. They are an important resource for sustainable construction. They are appropriate for small-scale regional development in which local materials and skills are an essential component. It is therefore wrong to omit grey chalk lime from the British and European Standard on building lime, which should be revised accordingly.

This paper was submitted to the relevant British and European Standards Committees in 2006 with a request that the reintroduction of grey chalk and slightly hydraulic limes should be included in the forthcoming review of BS EN 459 for building limes. Unfortunately this request was turned down for the 2010 revision of the Standard. A principal reason given for this refusal was that the compressive strengths stated in BS 890, from 0.7 to 2.0 N/mm^2 at 28 days, could not be achieved with the test method now mandated for the stronger NHLs. Clearly, to enable continued use of grey chalk limes, appropriate test methods must be reinstated before formal recognition of these limes can be possible in future editions of the Standard.

This paper was first published in Volume 13 of the *Journal of the Building Limes Forum* in 2006. This is a revised and updated version.

Notes

1   Smeaton, J., *A Narrative of Building and a Description of the Construction of the Eddystone Lighthouse with Stone*, 2nd edition, G. Nicol, London, 1793, p. 115.

2   Vicat, L. J., *Mortars and Cements*, trans by Smith, 1837, reprinted by Donhead Publishing Ltd, Shaftesbury, 1997.

3   Cowper, A. D., *Lime and Lime Mortars*, BRE, London, 1927, reprinted by Donhead Publishing Ltd, Shaftesbury, 1998.

4 British Standards Institution (BSI), *BS 890:1972. Specification for Building Limes*, BSI, London, 1972, p. 7.

5 Everett, A., revised Barritt, C. *Mitchell's Materials*, 5th edition, Longman Scientific and Technical, Harlow, UK, 1994, p. 112.

6 Acknowledgements and thanks to Michael Wingate for this information from *The Uses of Lime in Building*, published by The Southern Lime Association.

7 Salzman, L. F., *Building in England down to 1540*, Clarendon Press, Oxford, 1952, p. 83.

8 Salzman, ibid. p. 125.

9 North, F. J., *Limestones: Their Origins, Distribution and Uses*, Thomas Murby & Co., London and New York, 1930, p. 398.

10 Smeaton, *op. cit.*

11 Smeaton, *op. cit.*, p. 116.

12 Smeaton, *op. cit.*, p. 117.

13 North, *op. cit.*, p. 393.

14 Dobson, E., *Rudiments of the Art of Building*, John Weale, London, 1854, section I, p. 70.

15 Burnell, G. R., *Rudimentary Treatise on Limes, Cements, Mortars, Concretes, Mastics, Plastering etc*, 6th edition, Virtue & Co., London, 1867, p. 51.

16 Burn, R. S., *The New Guide to Masonry, Bricklaying and Plastering*, John G. Murdoch, London, 1871, p. 43, reprinted by Donhead Publishing Ltd, Shaftesbury, 2001.

17 Gwilt, J., *An Encyclopaedia of Architecture*, revised by Wyatt Papworth, J. B., Longmans Green & Co., London and New York, 1894, pp. 532–3.

18 Rivington, N., *Series of Notes on Building Construction* Part III *Materials*, 4th edition, Longmans Green & Co., London, New York and Bombay, 1899, p. 160, reprinted as *Rivington's Building Construction*, Donhead Publishing Ltd, Shaftesbury, 2003.

19 Rivington, ibid, p. 154.

20 Caxton House Editorial, *Specification*, incorporating the Municipal Engineer's Specification, No. 11, London, 1908–9, p. 183.

21 Caxton House Editorial, ibid, p. 230.

22 Caxton House Editorial, ibid, p. 240.

23 Dibdin, W. J., *The Composition and Strength of Mortars*, The Royal Institute of British Architects, London, 1911.

24 Howe, J. A., *The Geology of Building Stones*, Edward Arnold, London, 1910, reprinted by Donhead Publishing Ltd., Shaftesbury, 2001, p. 183.

25 Cowper, *op. cit.*, pp. 24, 58.

26 Searle, A. B., *Limestone and its Products*, Ernest Benn Limited, London, 1935, p. 24.

27 Searle, ibid, p. 570.

28 McKay, W. B., *Building Construction*, Longmans, Green and Co., London, 1938–1952, reprinted by Donhead Publishing Ltd., Shaftesbury, 2005, p. 20.

29 British Standards Institution (BSI), *BS 890:1940. Specification for Building Limes*, BSI, London, 1940.

30 British Standards Institution (BSI), *BS 890:1966. Specification for Building Limes*, BSI, London, 1966.

31 BS 890:1972, *op. cit.*

32 Everett, A., *Mitchell's Building Construction*, B. T. Batsford Limited, London, 1978, p. 141.

33 Everett, 1994, *op. cit.*

34 Smith, W., Original Geological Map, displayed at Burlington House, London, 1815–2006. (Winchester, S., *The Map that Changed the World*, Viking, London, 2001.) Geological Maps – Maps showing excellent geological information including the location of limestone formations to a range of scales and detail may be obtained from the British Geological Survey, at the BGS Sales Offices in Nottingham, Edinburgh and London. Geological maps are also available from Ordnance Survey agents and the HMSO Publications Centre or through booksellers.

35 North, *op. cit.*, p. 251.

36 North, *op. cit.*, p. 435.

37 Vicat, *op. cit.*

38 Holmes, S. and Wingate, M., *Building with Lime*, Intermediate Technology Publications, 1997, p. 13.

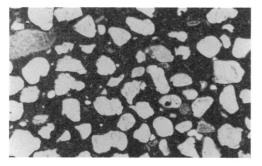

Plate 1 1930s hydraulic lime:sand render containing uniformly graded quartz sand with rounded to sub-rounded grains from Ryder Seed Hall, Hertfordshire.
Field of view = 4.5 mm across.

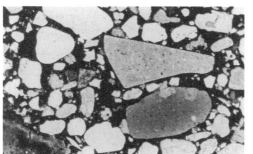

Plate 2 1930s hydraulic lime:sand render containing well graded Thames valley river terrace sand. Sample from Ryder Seed Hall, Hertfordshire. Aggregate chiefly comprises quartz (white) and flint (light brown).
Field of view = 4.5 mm across.

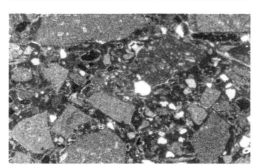

Plate 3 Crushed flint (grey) fine aggregate within eighteenth-century lime:sand render from south-east England. An unburnt chalk particle (brown) indicates that the lime was manufactured from the Cretaceous Chalk.
Field of view = 4.5 mm across.

Plate 4 Granite derived fine aggregate within a recent mortar from south-west England. Comprising sub-angular particles of quartz (white/grey), feldspar (grey) and including flakes of muscovite mica (brightly coloured).
Field of view = 2.5 mm across.

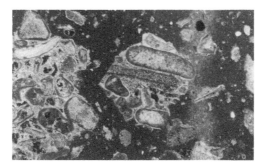

Plate 5 Crushed Jurassic limestone fine aggregate within an eighteenth-century lime:sand render from Thame Park House, Oxfordshire.
Field of view = 4.5 mm across.

Plate 6 Aggregate chiefly comprising unfossilised shell fragments within nineteenth-century lime : sand render from South Bishop Lighthouse, Pembrokeshire. Shell fragments include bivalves and echinoderms.
Field of view = 4.5 mm across.

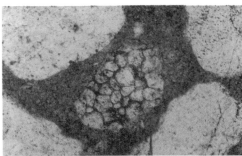

Plate 7 Relict grain comprising dicalcium silicate (belite, centre) within a 1930s hydraulic lime render. From Ryder Seed Hall, Hertfordshire.
Field of view = 0.5 mm across.

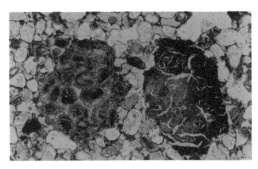

Plate 8 Inclusions of unburnt limestone (left) and an overburnt lime lump (right) within Late Medieval lime:sand mortar from Oystermouth Castle, Gower.
Field of view = 4.5 mm across.

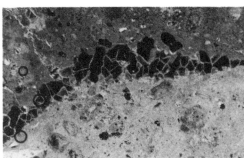

Plate 9 Partially burnt limestone inclusion within first-century lime mortar from Piddington Roman Villa, Northamptonshire. Showing the burnt particle rim around unburnt core (lower half).
Field of view = 4.5 mm across.

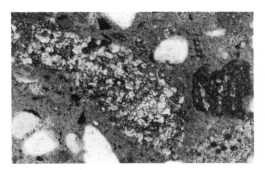

Plate 10 Large relict Portland cement grain (centre) within an early twentieth-century lime : Portland cement : sand render from Torrington House, Hertfordshire.
Field of view = 1.0 mm across.

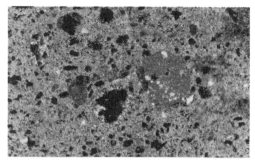

Plate 11 Gypsum crystals lining air void (grey) and clay impurities (orange) from the gypsum addition within an early twentieth-century lime plaster gauged with gypsum, from Torrington House, Hertfordshire.
Field of view = 2.5 mm across.

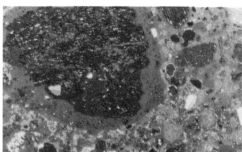

Plate 12 Pumice aggregate (upper left) with pozzolanic reaction rim at the interface with the lime binder. Mortar from the Roman concrete harbour structure at Corinth, Greece.
Field of view = 4.5 mm across.

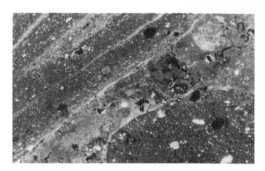

Plate 13 Ceramic tile (red) used as aggregate within first-century lime mortar from Piddington Roman Villa, Northamptonshire.
Field of view = 4.5 mm across.

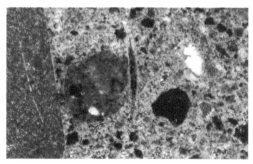

Plate 14 Brick dust inclusions (red) within an eighteenth-century lime:sand mortar from Hertfordshire.
Field of view = 1.0 mm across.

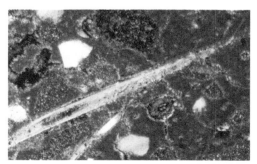

Plate 15 Animal hair (yellow) reinforcement in lime:sand plaster from a nineteenth-century school building in Hertfordshire.
Field of view = 1.0 mm across.

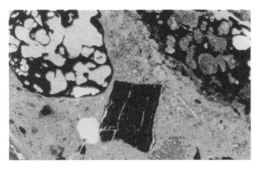

Plate 16 Inclusions of charcoal (black) within Napoleonic lime:sand mortar from Fort Clarence, Rochester.
Field of view = 2.5 mm across.

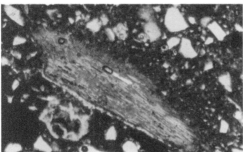

Plate 17 Vegetable matter inclusion within an eighteenth-century lime:sand mortar from Hertfordshire.
Field of view = 1.0 mm across.

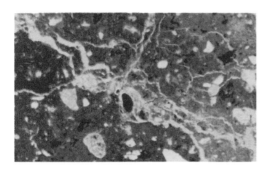

Plate 18 Cracks filled with gypsum within lime:sand brickwork mortar from Buckinghamshire.
Field of view = 1.0 mm across.

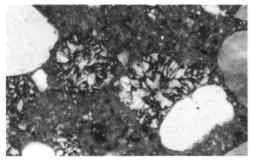

Plate 19 Secondary deposits of the mineral ettringite (grey) filling air voids of a Portland cement:sand mortar from Earl's Court Underground Station façade, London.
Field of view = 1.0 mm across.

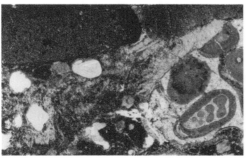

Plate 20 Replacement of binder by thaumasite (yellow) within a Portland cement:lime brickwork mortar from a basement in Oxford.
Field of view = 4.5 mm across.

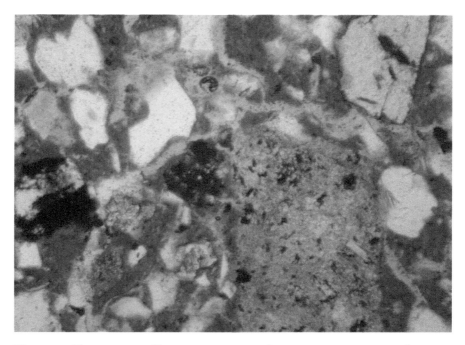

Plate 21 Thin section of lime-putty mortar, showing an interconnected pore structure in blue. (B. Revie, Construction Materials Consultants C.M.C.)

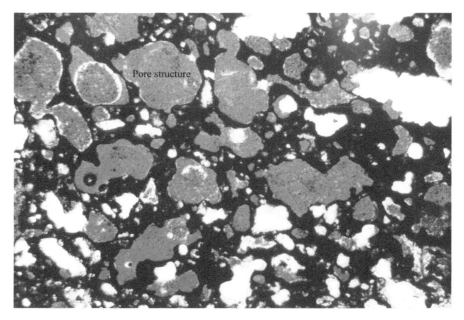

Plate 22 Thin section of hot-lime mortar, showing a relatively cavernous interconnected pore structure in blue. (Scottish Lime Centre Trust)

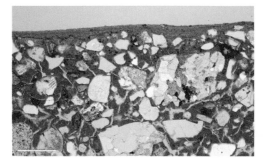

Plate 23 Hot mortar. Coarse coat with lime–sand ratio 1:2 (vol). Fine coat 1:1.5 (vol), compacted. Thin section of a sample from eastern façade (2005) (T. Seir Hansen).

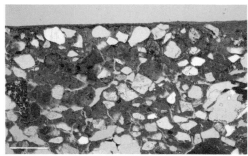

Plate 24 Wet mortar. Coarse coat with lime–sand ratio 1:2 (vol). Fine coat 1:1,5 (vol), compacted. Thin section of a sample from eastern façade (2005) (T. Seir Hansen).

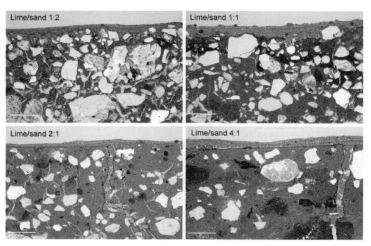

Lime/sand 1:2
Lime/sand 1:1
Lime/sand 2:1
Lime/sand 4:1

Overflade

Plates 25–28 Experimentally reproduced historic mortars with different lime–sand ratios. Hot mortar samples (2005). First sample (lime–sand 1:2) represents the mortar used in restoration today. The last sample (lime–sand 4:1) has a structure quite similar to the original historic mortar as shown in Plate 29 below. Lime paint seals the shrinkage fissures (T. Seir Hansen).

Plate 29 Historic mortar from the original structure; seventeenth century (lime–sand ratio approx. 4:1). B = binder; L = air; S = sand; K = Lime inclusions. (T. Seir Hansen)

Plate 30 Thin section with different binder related particles ('wet mortar', lime/sand ratio 1:2). K = Lime inclusions, Kh = particles with hydraulic components, C = particles of incompletely burned lime, Gips = gypsum. Black fields = air pores (T. Seir Hansen).

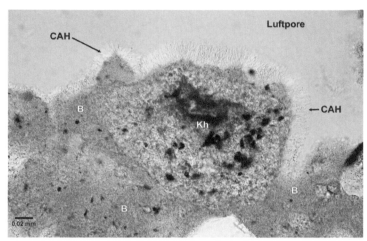

Plate 31 Thin section showing binder, B, ('wet mortar', lime/sand ratio 1:1) with a particle with hydraulic components, Kh . Needle-like crystals, CAH, are formed on the surface through hydraulic reaction with water. Yellow field = air pore (T. Seir Hansen).

Plate 32 Diffusion of nano-lime (E-25) through Ketton limestone. The maximum penetration is shown by pink phenolphthalein.

Consolidation of flaking paint	Consolidation of plaster and hairline fractures with nano-lime (E50)
Injection of fine lime-based grout material	Consolidation of plaster and hairline fractures with nano-lime (E-25)
Lime-based repairs	

Plate 33 Areas of repair and repair types to the wall painting of St George located in the nave of All Saints' Church, Little Kimble, on the north wall between two windows to the east of the door.

Rediscovering Methods and Techniques

Lime Mortars
The Myth in the Mix

Gerard Lynch

INTRODUCTION

In the past, the master mortar maker was heir to a long tradition of technical knowledge, understanding of local materials and pragmatic experience. He knew exactly how to prepare lime and aggregate – usually sand – in order to achieve the best results for the craftsmen masons, bricklayers or plasterers. They in turn, since the days of their apprenticeships, had developed an innate knowledge of these materials and a comprehensive grounding in their correct use. Thus, in the days before any legislative standards or guarantees as to material performance, they maximized the capabilities of the limes and sands at their disposal.

One of the crucial factors in preparing any lime-based mortar was, and remains, recognizing the capabilities of the lime as the 'binder', and identifying and establishing the nature of the sand as the 'filler'. This is in order to determine the best ratio of lime to sand, select the most appropriate and pragmatic method of their preparation (to gain maximum workability of the fresh mortar) and, once hardened, to achieve a final strength linked to suitability of purpose. In this paper references to 'ratios' will be, unless specifically stated otherwise, to proportions of constituents by volume.

HISTORIC AND MODERN MORTAR RATIOS

Analysis of historic mortars reveals that the types of limes and sands, and their mix ratios, varied considerably. Richard Neve illustrates this with examples of varying mortar ratios commonly used in and around London.[1]

These were normally different in various parts of the same building for the footings and inner and outer flank walls, with the best, lime-rich mortar, reserved for the outer leaf of the façade. To a large degree the type of lime and sand, and the need to obtain a workable mix, determined these ratios. It is, however, interesting to note that, with the lime revival of the past 25 years – which for many years was primarily based on the use of pure, non-hydraulic lime putty mixed with a well-graded aggregate – has seen an emphasis on the common use of a 1:3 lime:sand ratio that is based essentially on the 'voids by volume' within a measure of dry sand.

This is a correct method, and is based on the need to provide sufficient lime binder to ensure a coating of lime around every grain of sand and the voids between them. The test to determine void volume by this method, and thus the mortar ratio too, involves half-filling a graduated laboratory flask with an oven-dried sample of the specified sand, and another identical graduated flask half-filled with clean (potable) water. This water is then carefully poured onto the sand until it is floated level with the top of the sand. The amount of water necessary to fully saturate the sand is recorded by measuring the volume poured from the flask: this is taken to determine the volume of lime binder required for producing a good mortar. Typically this is found to be one-third of the original volume of the water, and hence the ratio is determined as 1:3.

DRY OR SAND-SLAKING

It would be incorrect, however, to believe that this was the only method by which the 1:3 ratio was historically specified. To understand this it is vital to recognize that the vast majority of quicklimes were not slaked to a putty, and subsequently gauged with the sand to make traditional mortars. Matured lime putty was normally only used by plasterers for internal plasterwork, and by bricklayers setting very finely jointed gauged brickwork. For general building work the quicklimes were simply slaked to a crude powder, or dry hydrate.

One of the most popular methods to achieve this was to slake a one-third measure of quicklime, broken down to the size of nutmegs, and place it within a ring of a cubic yard of sand. The minimum of water was added, and then the sand was quickly drawn over the heap as the lime both heated and broke down. This method was commonly referred to as sand or dry slaking. If the mortar was required for the general brickwork on side elevations, for backing up facework, internal walling or for footing

Figure 1a Quicklime within damp sand, prior to the addition of water, and the sand drawn over to cover and facilitate slaking within the sand heap.

Figure 1b The heap has now dried out, expanded and cracked open after the quicklime, in slaking, has increased in both temperature and volume.

courses, it would be turned over to fully mix the dry sand and lime once slaking was completed, and the mix cooled. Sufficient extra water would then be added to bring the resulting mortar to the desired working consistency, and it was then used immediately.

Alternatively, a similar mix could be used to produce a top quality front mortar, particularly favoured on façade masonry. In this case, once the lime was slaked and the mortar turned over, it could be thrown, or punched, by shovel through a large inclined 5 mm (¼ in.) meshed screen to remove large inclusions, such as particles of underburnt and overburnt lime, then mixed with water and left as a covered heap to stand, or bank, for a few days to fatten up. This would increase workability for knocking up, or reworking, to the desired consistency immediately before use.

It is important to note that neither of these describes a 'hot mix', as sometimes referred to by contemporary writers. A hot mix is only the correct term when a mortar is actually used immediately, while still hot from the slaking of the quicklime with sand and water.

QUICKLIME AND HISTORIC MIXES

The important aspect to note here is that the mortar, though mixed to the frequently quoted ratio of 1:3 lime:sand, was not mixed from prepared slaked lime or hydrated lime (calcium hydroxide), either as a putty or a powdered dry hydrate, but directly from the quicklime.[2] This vital point has frequently been overlooked, and has led to the misinterpretation of a great many historical mortar mixes based either on original site records of mortar ratios, or on mixes suggested within old craft books. It is important to recognize that up until the Second World War (1939–45) a majority of limes were still prepared from freshly burnt quicklime delivered to site, as opposed to ready-to-use slaked putties or bagged dry hydrates. Therefore, whenever one reads historically of a 1:3 ratio, the writer is usually taking it for granted that the reader would appreciate that he is referring to the quicklime:sand ratio, rather than one based on separately prepared lime. A simple yet good example of this is to be found in an architect's private site book, for an entry dated 1927, on preparing lime mortar as follows:

Mortar: Lime 1 Sand 3

Lime slack [slake] with water & then cover with sand. After lime is thoroughly slack screen through upright screen & then mix with water to desired consistency.

The writer is very familiar with, and frequently demonstrates, this process of dry or sand-slaking using both non-hydraulic and hydraulic quick-limes, alongside the better-known method of slaking to a putty. It is important to understand that all quicklimes increase in volume when they are slaked, so that the resultant lime:sand ratio for the finished mortar is always more lime-rich than the original stated ratio. The actual degree of increase varies with the type and class of lime, but is typically between 60% and 100%, or even more.

This is the reason that, under analysis, the majority of historic lime mortars are not commonly found to be 1:3 finished mixes, but typically vary between 1:1½ to 1:2; just as the original mortar makers and the craftsmen intended. This is borne out by the experience of the Scottish Lime Centre Trust, who at their last count had analysed around 4,500 historic mortar samples. Approximately 80% of these were samples from Scotland, 10% from England, and the remaining 10% from various other countries. They estimate that the lime:sand ratios of all samples on their database average around 1:1½.

This 1:3 ratio of quicklime to aggregate suited both non-hydraulic and hydraulic limes, most general building sands and most methods of slaking, and it particularly suited the feebly hydraulic class, historically termed stone, greystone or grey lime. Sometimes, however, particularly when transportation of materials was both difficult and expensive, builders had no choice except to use a poorly graded and/or excessively fine local sand, rather than the ideal well-graded building sand so frequently emphasized today. Such sands demand an increased lime content to make

Figure 2   Top centre: a sample of 'screened', dry-slaked 1:3 quicklime:sand mortar with, below it in the centre, a fully mixed and matured sample of a matching mortar, but now made from a 1:2 ratio of slaked lime:sand. On either side are four samples of historic mortars from the seventeenth and eighteenth centuries for comparison.

a good mortar, and the craftsmen simply revised the quicklime content accordingly to ensure a perfectly acceptable mortar.

The Scottish Lime Centre Trust undertook detailed analysis of several samples of original mortar during archaeological works to the external brick fabric of Aspley House in Bedfordshire (*c.* 1692; 'double-piled' in 1745). This was found to have been made using the fine sand obtained from within part of the curtilage of the property, mixed with the local (Totternhoe) feebly hydraulic grey chalk (stone) lime. These mortars had been used both for the mansion brickwork, and on a long and very tall garden boundary wall to the rear of the property. The main house brickwork mortars, from both phases of construction, were to identical ratios of 1 : 1.4. Interestingly the mortar for the very long, tall brick garden wall was to a ratio of 1 : 0.7, revealing that the bricklayers had doubled the ratio of lime to sand as a logical and pragmatic way to gain the additional strength and weathering capabilities deemed necessary for this much more exposed element.

The process of dry or sand-slaking lime mortars leaves very small particles of lime dispersed throughout the mortar, giving it an appearance on very close examination which is rather different from that of a putty-based mortar. This is because the process of maturing a putty allows the lime

Table 1   Conversion of Millar's recipes from Imperial to Metric. The recipes show the quantities of lime, sand, hair and water required per superficial yard (1 sq yd = 0.77 m²) for various classes of plastering.

Render Type	Material	Imperial	Metric	Parts by Volume
Render ⅜" thick (0.95 cm)	Lime (unslaked)	0.15 cu ft	8.1 litres (lime putty)	1.25
	Sand	0.23 cu ft	6.5 litres	1
	Hair	0.1 lb	0.05 kg	
	Water	1.2 gal	5.4 litres	
Render + Set ½" thick (1.27 cm)	Lime (unslaked)	0.22 cu ft	11.9 litres (lime putty)	1.8
	Sand	0.23 cu ft	6.5 litres	1
	Hair	0.12 lb	0.05 kg	
	Water	1.8 gal	8.2 litres	
Render + Float ⅝" thick (1.59 cm)	Lime (unslaked)	0.25 cu ft	13.5 litres (lime putty)	1.25
	Sand	0.38 cu ft	10.8 litres	1
	Hair	0.17 lb	0.077 kg	
	Water	2.0 gal	9.1 litres	
Render + Float +Set ¾" thick (1.91 cm)	Lime (unslaked)	0.32 cu ft	17.2 litres (lime putty)	1.6
	Sand	0.38 cu ft	10.8 litres	1
	Hair	0.18 lb	0.082 kg	
	Water	2.6 gal	11.8 litres	
Setting: Putty + Gypsum ⅛" thick (0.3 cm)	Lime (unslaked)	0.1 cu ft	5.4 litres (lime putty)	6.4
	Gypsum	0.03 cu ft	0.85 litres	1
	Water	1.0 gal	4.55 litres	

particles to become progressively finer and finer over time. These particles are, however, equally visible in the majority of historic building mortars, adding weight to the idea that these were originally mixed by this method, and not from matured putty.

Concern has sometimes been expressed that these small particles are unslaked lime, and thus a defect in the mix which will spoil the work. However, any particles of unslaked lime from the initial slaking phase are re-activated after the lime and sand are dry mixed, when further water is added as the mortar is prepared and 'banked'. This allows a further period of slaking before use. Normally, all particles left after this and the knocking-up stages simply behave as aggregates within the mortar, and do no harm to the finished work.

Table 1 shows a table of lime render and plaster recipes from William Millar[3] modified by Paul D'Armada of Hirst Conservation to include metric conversions of the specified imperial quantities, and the equivalent volumetric mixes. The relative richness of these mixes is very striking.

LIME TYPES AND BULK DENSITIES

A lack of recognition of the traditional method of dry or sand-slaking has been a cause of failures in mortars based on a volume ratio of 1:3 with ready-to-use lime. Another contributing factor is inexperienced personnel working with lime putty who do not realize a measure of lime within a ratio might not be one full unit of lime, but contain a sizeable percentage of water; thus reducing the actual binder content within that ratio further. One must always discuss with the lime supplier the best method to achieve the specified volume ratio when lime putty is the specified binder. All limes come from different raw feedstocks, and as Paul Livesey states:[4]

> ... the quicklime (now termed CL Q, or Calcium Lime – Quicklime) will vary in characteristics, which is a major factor with bulk densities variable between 900 and 1200 kg/m³,[5] as is slaked lime (CL S, or Calcium Lime – Slaked) at 350 to 640 kg/m³ ... whereas relative density (formerly termed specific gravity) is the density of combined free water + suspended solids + dissolved solids, but excluding air, bulk density, on the other hand, includes the above, but also any entrapped air and is generally lower than the relative density. Therefore, bulk density is what one gets in the bag of dry hydrate or tub of lime putty; and is the value to use when calculating the 'lime' (CL S) content.

A good mature putty (four months old as opposed to fresh putty) will have a relative density of 1.450 kg/m³, weigh approximately 1.45 kg/litre and contain 640–650 g (equivalent dry weight) of lime per litre, or 470–480 g/kg.

Non-hydraulic and hydraulic limes are both now available as dry hydrates. The former as High-Calcium lime, which is usually marketed as builders' lime, and primarily intended as a plasticizer in cement:lime:sand mortars (1:1:4 or 1:1:6 etc) for modern masonry construction. This processed lime is not, however, a good substitute for a traditional non-hydraulic lime putty, or for use on traditionally constructed buildings, as it does not posses the same working characteristics as traditionally slaked lime putty. It is not intended for use in lime:sand mortars, and would therefore not meet the required strength and durability performances.

Modern dry-hydrated hydraulic limes, marketed as Natural Hydraulic Limes (NHLs), are classified in three ascending numerical grades of compressive strength at 28 days, expressed in N/mm^2, as NHL2, NHL3.5 and NHL5; which are broadly equivalent to, but not parallel to, the old classifications of feebly, moderately and eminently hydraulic limes respectively. When gauging NHLs with sand to make a mortar it is important to understand that dry hydrates have different bulk densities to sand (as do all powder binders) and therefore should ideally be mixed by weight, not volume. As weigh batching is rarely practised on site, most lime suppliers specify volumes of sand (usually to the nearest 10 litres) per full bag of NHL (weight varies).

It is also important to remember that damp sand increases, or bulks, in volume (the amount being dependent on sand grading and moisture content), whereas saturated and bone dry sand have identical volumes, so an allowance also has to be made for this when measuring the sand. Only then can the lime and sand be accurately volume batched to each other to the specified ratio. Again, it is important to discuss this, and to agree the correct procedure with the lime supplier.

This article was first published in *The Building Conservation Directory* in 2007 and is reprinted here with the kind permission of the publisher, Cathedral Communications Ltd. It was subsequently republished in the *Journal of the Building Limes Forum*, Volume 16, 2009, and has been further revised and expanded by Dr Lynch for this publication.

Notes

1 Neve, R., *The City and Country Purchaser and Builders Dictionary*, Sprint, Rivington and others, London, 1736.
2 Lynch, G. C. J., 'Lime mortars for brickwork: Traditional practice and modern misconceptions', *Journal of Architectural Conservation*, Vol. 4, No. 1 and 2, March and July 1998.
3 Millar, W., *Plastering Plain and Decorative*, 1897, reprinted by Donhead Publishing Ltd, Shaftesbury, 1998.
4 Livesey, P., in private correspondence with the author, 2011.
5 Oates, J. A. H., *Lime and Limestone: Chemistry and Technology, Production and Uses*, Wiley VCH, Weinheim, 1998.

The Use of Pozzolans in Lime Mortars

Geoffrey Boffey, Elizabeth Hirst and Paul Livesey

INTRODUCTION

In the conservation and restoration of old buildings, a great deal of time and effort is spent on matching old materials (perhaps centuries old) using modern chemicals and products. Consideration has also to be given to the techniques and skills of the old craftsmen, which in most cases reflected the materials available at the time. Modern materials are much more tightly specified and react much more predictably than the materials manufactured 100 years ago and while modern craftsmen are still highly skilled, there is a higher expectation both of predicted performance and economics than there was centuries ago. Materials have also changed and high calcium lime is typical of such a material. Over the years, this has changed from being an impure raw material into a tightly controlled 'chemical' due to pressure, first, from the chemical industry and second, from the steel industry. The same changes have not been so apparent in the materials used as pozzolans. While efforts have been, and are being made, to characterize the effects of some of the traditional pozzolans, newer materials are being considered because of their pozzolanic effects. Although some British and European Standards for these have been produced, there is very little specific documented information regarding their properties and performance characteristics when used with lime. The Donhead publication, *Hydraulic Lime Mortar*, does start to give some guidance.[1]

The fact that the properties of lime mortar can be modified by the addition of other materials has been known for some time. The earliest recognition of this seems to have been in Mediterranean countries where volcanic material is readily available in the form of volcanic ash, tuff or pumice, and its widespread use around Pozzuoli in Italy led to the name pozzolan or pozzolana being used to describe such materials. Pozzolanic materials do not harden in themselves when mixed with water but, when finely ground and in the presence of water, they react at normal ambient temperature with dissolved calcium hydroxide $(Ca(OH)_2)$ to form strength-developing calcium silicate and calcium aluminate compounds. These compounds are similar to those which are formed in the hardening of hydraulic materials. Pozzolanas consist essentially of reactive silicon dioxide (SiO_2) and aluminium oxide (Al_2O_3), the remainder contains iron oxide (Fe_2O_3) and other oxides while for most their content of reactive calcium oxide is negligible. For true pozzolanic reaction the silicon dioxide content should be not less than 25.0% by mass. The situation is further complicated by the lack of definitive information about when and how such materials should be used.

PERFORMANCE AND CLASSIFICATION OF LIME/POZZOLANS

In conservation, the two types of lime used are high calcium (non-hydraulic) lime and natural hydraulic lime (NHL). There are essential differences between these types of lime, and their behaviour with pozzolans can also be different.

In the United Kingdom, high calcium lime is predominantly a high purity material produced from the calcination of high purity limestone. Changes in production techniques involving more efficient means of cleaning the stone prior to calcination, together with higher efficiency kilns, result in today's material being of greater purity and having a better consistency with predictable properties. The lime is much more reactive with water, and modern slaking techniques give an extremely good lime putty. When used with sand in a mortar, the initial setting process is mainly due to water loss, with the long-term hardening action being a result of carbonation.

With high calcium lime, a true pozzolanic additive will react with the lime in the presence of water to produce cementitious (primarily amorphous calcium silicate hydrates) products, which increase the binding of the sand as shown by an increase in compressive strength:

$Ca(OH)_2 + H_2O + SiO_2.Al_2O_3 \rightarrow$ hydrated calcium silicates

(Lime) (water) (Pozzolan) + hydrated calcium aluminates

Lime – pozzolan reaction

Amorphous alumina found in some pozzolans (e.g. pumice) will also react with water to form hydrated calcium aluminates.

This action may or may not increase the initial setting rate. Certain additives (principally porous materials) can, however, speed up the setting action by introducing entrained air (which includes CO_2 for carbonation of the free lime) and by improving capillary action for the removal of water. Some such additives are often incorrectly described as being pozzolanic.

NHL is, by comparison, produced by calcining impure limestone, usually at higher temperatures than for high calcium lime. At these higher temperatures, the silica, alumina and iron in the impurities react with the lime to give cementitious minerals that will react with water to give a set and ultimate hardening. NHLs have the property of hardening under water and are further classified as being feebly (NHL1–2), moderately (NHL3.5) or eminently (NHL5) hydraulic depending on their hardening performance.

A more scientific classification used, often to assess the potential of the limestone and it assumes all silica becomes activated, is the cementation index (CI). This is calculated from an analysis of the lime/limestone by giving different weightings to the various chemical components.[2]

$$CI = \frac{\text{Soluble Silica} + Al_2O_3 + Fe_2O_3}{\text{Total CaO} + MgO}$$

However, in general, natural hydraulic limes are classified according to the percentage of soluble silica they contain by mass, sometimes referred to as combined silica, i.e. that combined with the calcium oxide and aluminium oxides in calcium silicates and aluminate phases.

NHL1–2 contains 3–6% soluble silica, NHL3.5 contains 6–12% soluble silica and NHL5 contains 12–18%.

The setting and hardening of NHLs is a combination of the hydration of the cementitious components together with the normal carbonation reaction shown by high calcium limes. Although modern chemical methods have helped producers of NHLs to improve the consistency of their products, NHLs, relying on their natural raw materials, have a less precise uniformity of composition than high calcium limes and setting performance varies among producers. Pozzolanic additives react with the free lime present in the NHL and that released as the cementitious components hydrate. They usually have a much greater effect on a weakly hydraulic lime, rather than an eminently hydraulic lime, as there is usually more free lime available. Porous additives have a similar effect with hydraulic limes as with high calcium limes.

The term 'pozzolan' covers a wide range of materials containing reactive silica which can react with lime at ambient temperatures to form calcium silicates.[3] This definition covers both natural and synthetic materials (Table 1).

Table 1   Various natural and synthetic pozzolans.

Natural pozzolans	Synthetic pozzolans
Volcanic ash	Ground granulated blastfurnace iron slag
Tripoli	PFA (Fly ash or pulverized fuel ash). Some contain too much sulphate
Diatomaceous earth	Ground brick or tile (burnt at or below 900°C
Pumice dust	Kaolinite/Metakaolin (e.g. Metastar) – clay burnt at about 900°C
Trass	HTI (High temperature insulation – finely ground china clays)
Metamorphosed siliceous rocks containing soluble silica	Bauxite burnt at low temperatures – 200–300°C

The range of materials that can be classed as pozzolans is wide; and in India and other Asian countries, for example, it has been common practice to mix burnt clay with lime mortars. In the United States, low-iron slag or fly ash mixed with hydrated lime has been used extensively as a non-staining mortar for laying marble, cut limestone and polished masonry for flooring.[4]

Soils with a high clay content can even act as pozzolans, and this is used in soil stabilization where lime is added as a minor component. After mixing into the soil, followed by watering and rolling, it will set to give a firm base for future construction work. This requires a fairly high ambient temperature and is less frequently used in the United Kingdom than in warmer countries.

The degree of reactivity with lime varies tremendously and often there can be wide variations even within the same type of natural material from the same source. Overall, the reactivity depends on the chemical composition of the material (especially the amount of available reactive silica), the fineness of the pozzolan, and the reactivity of the lime. The speed of the reaction between the pozzolan and the lime will also depend on the temperature and there being water available during the reaction period (often several months).

In ancient times the pozzolan consisted of large fragments and dust and was generally mixed in ratios of 2–3 parts pozzolan to 1 part of lime putty. Since the larger parts of the pozzolan does not react with lime it acted as an inert filler or aggregate, reducing shrinkage and negating the use of sand (which is rarely found in Roman technology).

While it is possible to obtain reasonable uniformity in some synthetic pozzolans, it is often difficult to obtain materials with consistent composition and performance. European Standards have imposed specifications on the performance of such pozzolans as: Fly Ash (BS EN 450-1); Ground granulated slag (BS EN 15167); and Silica fume (BS EN 13263) while Metakaolin is part-way to a standard, having an Agrément Certificate, and there is an application for a standard for natural pozzolan.

In the United States a temporary specification (ASTM C432-59T) can possibly be used as a guide to pozzolan reactivity.[5] In the ASTM specification, the pozzolan is defined as a siliceous or alumino-siliceous material that in itself has little or no cementitious value, but that in finely divided form and in the presence of moisture will chemically react with lime at ordinary temperatures to form compounds possessing cementitious properties. To comply with this specification, pozzolans must conform to the following chemical and physical requirements:

- Moisture   10% max.
- Water soluble content   10% max.
- Loss of ignition   10% max.
- Fineness when wet sieved:

  * retained on 30 mesh (ASTM)   2% max. (590 μm sieve)
  * retained on 200 mesh (ASTM)   10% max. (74 μm sieve)

The specification also includes a lime-pozzolan compressive strength requirement (ASTM C114-61).[6] This uses a standard lime : sand : pozzolan mix with standard curing conditions plus a hydraulic compressive strength test that follows the same procedure, but with no lime and a higher pozzolan : sand ratio.

This specification was developed some time ago and may be more relevant to the production of manufactured mixtures than to the conservation industry, which is more often looking for particular solutions to specific application problems. The use of such a specification or something similar could, however, be a useful starting basis for the identification and broad screening of pozzolans for a particular application.

OTHER FACTORS

Historically, the use of lime-pozzolan mortar mixes disappeared in the United Kingdom after the Roman occupation and remained virtually forgotten. Smeaton, in the eighteenth century, discovered the pozzolanic-like properties when less-pure limestone was burnt to the correct temperature to form NHL. The advent of ordinary Portland cement in the late nineteenth century eventually resulted in NHL and lime-pozzolan mixes again virtually disappearing in the United Kingdom. Only in recent time has the need to restore and conserve our ancient buildings accorded to them a higher priority. The authors acknowledge and applaud the inclusion of brick dust in the work of the Smeaton Project, which in broad terms has so far concluded that a fine particle size of dust gives better results than coarser sizes, and dust from bricks fired at low temperatures gives better results than dust from those fired at higher temperatures.[7]

The detailing and characterization of material must also take into account other beneficial properties of a proposed pozzolan. Experience has shown that broken and powdered brick has the following properties:

1. acts as an anti-shrinkage agent;
2. is porous and will retain moisture within the mortar for a longer time enabling the lime carbonation crystal structure to form;
3. may provide active silica sites enabling the formation of calcium silicates and improved binding and adhesion of aggregates;
4. when partially dry, the improved porosity enables better and deeper penetration of carbon dioxide into the mass and thus greater depth of carbonation of the free lime component and hardening.

The particle size of the proposed pozzolan is also important. Roman plaster commonly contained broken pieces from 6 mm to dust and foundation concrete from Roman baths contains pieces from 25 mm down. Such mortars and concretes are potentially reusable and history shows that the Saxons reused Roman mortar and the Normans reused Saxon mortar by pounding it up and adding more lime. In this way, a lighter, even less dense and more porous mortar was produced. It is presumed, therefore, that the pounding process has exposed unreacted pozzolanic sites, which are available for subsequent lime reaction.

Even the type and particle size of the non-pozzolanic aggregate can have an effect on hardness, setting time and handling qualities of lime mortars. For example, stone aggregate inevitably produces a relationship in terms of temperature and weight between the mortar and the building stone. Temperature is well known as a factor that can affect the speed of setting and attainment of hardness of a lime mortar. Density and compaction of the aggregate is equally important, since it can be related to the speeds at which heat is absorbed (depending on specific heat and thermal conductance values for the material) and hence temperature stabilization within a mortar joint.

The temperature and density of aggregates are therefore interrelated and it is vital that the type of aggregate, whether stone or sand, should be stated for any mix. Usually, silica sand is dense and takes time to attain an equilibrium temperature with the stone. This can result in longer setting times and a denser mortar. In general, dense mortars inhibit evaporation and carbon dioxide transmission, whereas Roman and medieval mortars were designed to give easy and fairly rapid evaporation.

STANDARDS AND CHARACTERIZATION

The use of pozzolans in lime mortars can, and does, play an important part in restoring the fabric of ancient and historic buildings. It is very helpful to have those standardized pozzolans with well-defined properties so that, within reasonable parameters, lime mortars could be produced with more predictable properties.

The revised European Standard BS EN 459-1 divides limes into two families: Air limes and Hydraulic limes. Hydraulic limes are further subdivided into three types: Natural Hydraulic Lime (NHL); Formulated Lime (FL); and Hydraulic Lime (HL). Each of these is classified according to their compressive strength: Classes 2, 3.5 and 5. NHLs are not permitted any other constituent; FL, the previous NHL-Z, can have other constituents, pozzolans and even cement, but these must be declared; HL can have other constituents and these are not required to be declared, even cement. HL is often the 'economy' product used for such applications as ground improvement.

The allowable strength range might at first sight appear to be very broad; however, when the conformity statistical approach is taken into account the real ranges are much tighter with minimal overlap: NHL2–3.5 to 5.8; NHL3.5–5.6 to 8.3; and NHL5–8.2 to 12.5. In addition, the specified available lime and the allowable sulphur trioxide (SO_3) contents are tightened so that for SO_3, the requirement is less than 2% and only HL is relaxed to 7% SO_3 if soundness is demonstrated. This is another factor to consider concerning the suitability of HL for conservation work.

Overall, it is essential that the chosen product should be appropriate for the intended application and, while the European Standards may provide clarity, there is still an ongoing requirement for a better understanding of what properties pozzolans bring to lime mortars. Also the standards test the products as they leave the factory and therefore inappropriate storage in transit and at stockists can be detrimental to performance.

TESTS FOR POZZOLANICITY

Our own search for pozzolanic materials with defined properties identified a very highly reactive manufactured silica product (silica fume) that has a strong pozzolanic action with lime. Laboratory tests indicate that, with this pozzolan, an initial 'set' of a mortar is attained in about three days, with a high strength being achieved in about ten days. The prime merit of this material is that it is produced to a well defined chemical and physical specification. On this basis, its pozzolanic effect should be predictable and repeatable. Perhaps its only disadvantage is its dark grey colour, which will be imparted to the mortar. Other companies have worked along similar lines and evaluated pozzolans manufactured from clays and other siliceous materials.

With NHLs, pozzolans can be expected to react with the free lime while the cementitious components hydrate to give faster initial set with possibly a higher final strength. The European Standard characterizes NHL into three categories and gives their free lime content, together with compressive strength requirements. Within the conservation industry, there is considerable debate regarding the key properties of lime: pozzolans and whether or not higher compressive strength is desirable.

Screening and characterizing the various pozzolans is obviously an extensive task, and to this effect the authors have recently devised a simple chemical test that determines whether a material has pozzolanic activity. This test requires further refinement, but it is has already enabled comparisons of pozzolanic activity. The ultimate aim is to produce a table that should prove helpful when specifications are being prepared.

The principle of the test is that reactive silica (in a pozzolan) will combine with the lime in a saturated solution of limewater and reduce the alkalinity of the solution. By testing the pH of the solution at regular intervals the rate of pH reduction and therefore the rate of pozzolanic activity can be monitored.

Results:

Days	pH Values									
	Limewater (control)	Silica fume	Coarse Blastfurnace Slag	Fine Red brick dust	Fine Yellow brick dust	Fine Bath Stone dust	PFA	China clay	HTI	Fine grained Silica sand
0	12.7	12.7	12.7	12.7	12.7	12.7	12.7	12.7	12.7	12.7
6	12.7	11	12.7	11.5	11.5	12.7	12.7	12	12.7	12.7
11	12.7	10	12.7	10.5	10.5	12.5	12.5	11.5	12.7	12.7
40	12.7	10	12.7	10	10	12.5	12	11.5	12.7	12.7
60	12.7	10	12.7	10	10	12	11.5	11	12.7	12.5

These preliminary results suggest that:

1. Silica fume and brick dust (Bulmer low temperature fired red and yellow brick dust) react faster than the other materials tested, with silica fume reacting the fastest. Other laboratories have found that the reaction of Metakaolin is similar.
2. China clay, PFA (pulverized fuel ash) and Bath stone dust produced a slower reaction.
3. Silica sand gave no reaction, whereas HTI (high temperature insulation 'ceramic' type material) and coarse slag (iron blastfurnace slag) showed negligible or very slow reaction.
4. The results suggest that not all HTI materials give a pozzolanic reaction. The modern products have been subject to very high temperatures but possibly, as with bricks, the lower temperature produced types are more pozzolanic.
5. The slag was very coarsely graded with high volume to particle surface area ratio, which explains its low reactivity.

The standard BS EN 196-5 test for pozzolanicity of a pozzolanic cement is precise but requires a modest analytical laboratory as it is quite complicated involving titrations with EDTA: 20 g of the pozzolanic cement is mixed with 100 ml of boiled water and left for 8–15 days at 40°C to reach equilibrium. After this time the CaO and Hydroxyl ion concentrations in the solution are determined. If the concentration of CaO in this solution is less than that required to saturate a solution of the same alkalinity then the pozzolan in the cement is considered to be effective.

The more reactive the pozzolan in the cement the greater the amount of CaO required to saturate this solution, i.e. the difference between the CaO required for saturation of the test solution and the CaO already dissolved in it is the amount of CaO chemically combined with the pozzolan.

ADVERSE REACTION OF HYDRAULIC LIMES AND LIME-POZZOLANS

Today we are more aware of the risk of thaumasite and/or ettringite formation when hydraulic lime or pozzolanic limes are used to repair historic fabric containing gypsum or high concentrations of sulfates.[8]

Thaumasite ($CaCO_3.CaSO_4.CaSiO_3.15H_2O$) may be formed in persistently damp, cool materials which allow the cementitious compounds in hydraulic lime and lime–pozzolan mixtures to react with gypsum ($CaSO_4$) or other sulfates. Thaumasite is severely damaging to the fabric and converts hardened cementitious mortars to a pulp because of loss of strength and cohesion. It is less common than traditional sulfate attack as the formation of ettringite is postulated as an intermediate stage.

Ettringite ($3CaO.Al_2O_3.3CaSO_4.32H_2O$) is formed in similar conditions, but is mainly associated with cement, or when calcium aluminates are present in the hydraulic material, rather than hydraulic limes or lime–pozzolans. Its damaging effects are mainly manifest as cracking and spalling due to its expansion during formation (note its degree of hydration – $32H_2O$).

Consequently, to predict the long-term effect of a hydraulic lime or pozzolanic repair mortar, in the presence of gypsum and other sulfates, its compatibility with historic fabric should be assessed according to the Anstett test where the binder (cement, hydraulic lime or lime pozzolan mixture) is mixed to paste with 50% water by mass.[9]

After two weeks the hardened binder paste is crushed to 50 mm size, dried at 40°C and then mixed with gypsum (50% by mass of the dried binder paste) and ground to zero residue on a 900 mesh size.

This ground material is gauged with 6% water and compressed for 1 minute at 2 MPa in a cylindrical mould. The specimen taken from the mould is then placed on filter paper where the moisture is maintained by dipping its ends in water and covering the test with a glass plate to form an airtight seal.

If the diameter of the specimen does not expand by more than 1.25% after 90 days then the binder is safe to use for repairs where gypsum or other sulfates may be present.

Some pozzolans themselves may contain materials deleterious to historic fabric, e.g. salts such as sulfates in seawater quenched fuel ash or slag material, and should not be used.

While pozzolans are very useful given their setting properties with lime, they can also be chosen to match the colour of an existing mortar, if desired.

CONCLUSIONS

It is clear that there is a wide range of materials that can and do give a pozzolanic effect. The question therefore arises as to how the identification of such materials and the properties they bring to lime mortar performance is to be achieved. Evaluation and documentation is both daunting and expensive. While investigations have been carried out by sponsored organizations and individual companies, it is difficult to see where the financing of a wide-ranging test programme could come from in today's economic climate. An equally important issue is the practicality of using such materials on site. Although, in general terms, lime putty is consistent, not all producers apply the same control standards and variations can occur. A precise pozzolanic additive with specific mixing requirements may well cause on-site problems and some flexibility of addition rates is perhaps desirable. Fortunately, most pozzolan-lime reactions are relatively slow and a degree of tolerance is provided. With natural hydraulic lime, the faster setting and final strength achieved might, however, give a less tolerant mixture and may produce its own problems of long-term effects such as shrinkage and cracking.

There are therefore a wide range of parameters to be considered, and this paper has been written primarily to stimulate further debate and seek a wider interchange of information from specifiers, producers and users of those materials which, when added to lime or NHL, give a true or apparent pozzolanic action. A basic weakness within our industry is in describing a mortar mix in terms of lime, sand and pozzolanic materials, with specifications for individual materials in terms of separate performance, not that of the overall combination. Closer cooperation and an interchange of information can only result in improvement.

This paper was first published in the *Journal of Architectural Conservation*, Volume 5, No. 3, November 1999. This is a revised and updated version.

Notes

1 Allen, G., Allen, J., Elton, N., Farey, M., Holmes, S., Livesey, P., and Radonjic, M., *Hydraulic Lime Mortar for Stone, Brick and Block Masonry*, Donhead Publishing Ltd, Shaftesbury, 2003.
2 Boynton, R. S., *Chemistry and Technology of Lime and Limestone*, John Wiley & Sons, New York, 1966.
3 Ibid.
4 Ibid.
5 American Society for Testing and Materials, ASTM C432-59T introduced in 1959, discontinued in 1967 and replaced by C593 (*Standard specification for Fly Ash and other Pozzolans for Use with Lime*) in 1995.
6 American Society for Testing Materials, ASTM C114-61, updated as C114-99 (*Standard Test Methods for Chemicals Analysis of Hydraulic Lime*).
7 Ashall, G., Butlin, R. N., Teutonico, J. M. and Martin, W., 'Development of lime mortar formulations for use in historic buildings', *Durability of Building Materials and Components 7 – Proceedings of the Seventh International Conference on Durability of Building Materials and Components*, ed. Sjöström, C., Stockholm, 19–23 May, E. & F. N. Spon, London, 1966.
8 Collepardi, M., 'Thaumasite formation and deterioration in historic buildings', *Cement and Concrete Composites*, Vol. 21, No. 2, 1999, pp. 147–54.
9 Talero, R., 'Kinetochemical and morphological differentiation of ettringites by the Le Chatelier–Anstett test', *Cement and Concrete Research*, Vol. 32, 2002, pp. 707–17.

The Use of Lime Mortars and Renders in Extreme Weather Conditions

Paul Livesey

INTRODUCTION

Lime mortars and renders have been successfully used in the British Isles for many centuries. In fact the temperate climate in the UK is ideal for them, with a mean temperature at the optimum for hydraulic strength and the relatively high humidity levels contributing to hydration and carbonation. There are variations around such mean levels but, properly designed, placed and cured lime mortars and renders are capable of withstanding most extremes. However, these materials are particularly vulnerable during their early placing and maturing periods and this article is designed to promote an understanding of the threat mechanisms, the weather risk periods and the protection methods available to ensure successful installation.

PHYSICAL EFFECTS

Temperature and humidity

The setting and hardening properties of lime are chemical processes. Their action, in addition to the reactivity (class of lime hydraulicity) and concentration (mix design, lime and water content), is governed by thermodynamics, i.e. the temperature and relative humidity.

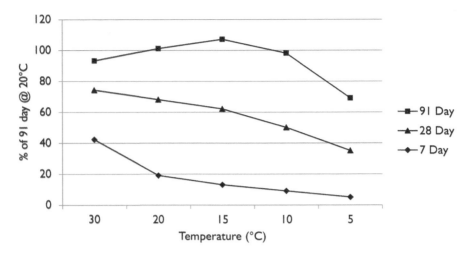

Figure 1   Effect of temperature on strength of NHL3.5 mortar.

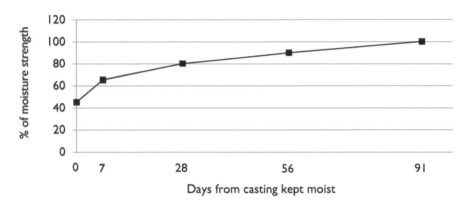

Figure 2   Effect of moisture curing period.

The effect of temperature is complex. Increased temperature results in higher initial action, reducing setting time and increasing early strength. It causes a dense phase to form on the surface of lime grains inhibiting further reaction and therefore resulting in lower final strengths (Figure 1). Lower early temperatures result in slower initial action, longer setting times and lower early strengths. However, a slight reduction in temperature produces a more porous hydration product and higher later strengths. In the case of much reduced temperature the ultimate strength might never be achieved. The optimum curing temperature is 15°C. This will have consequences for the build-up of resistance to aggressive forces.

Failure to maintain adequate humidity during curing will significantly reduce the ultimate strength. Leaving work exposed to atmospheric conditions can result in a loss of 50% of strength (Figure 2). Protecting for only seven days will lose up to 30% of strength.

The combined effect of temperature and curing humidity is that high temperature combined with drying conditions and low temperature combined with high humidity (saturation) present the most risk of failure.

Action of frost

The general action is well known. Ice has a density of 92% of that of water, hence ice volume will be 108% of the volume of the unfrozen water so that damage results from swelling on freezing. In practice, however, the matter is even more complex.

In the case of a capillary cavity partially filled with water (Figure 3), on freezing there is space for the extra volume of ice so no damage results. In the case of saturated mortar, where a capillary cavity is fully filled with water (Figure 4), either there has to be pore space into which the water can be pressed or the cavity will develop hydraulic forces that crack the surrounding mortar.

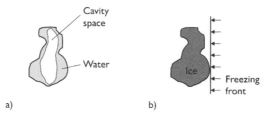

Figure 3   A simple partially filled cavity has space to expand without damage.

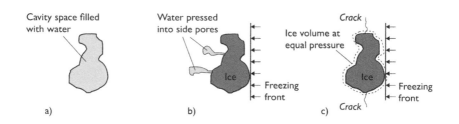

Figure 4   A simple cavity completely filled with water will cause expansive pressure on surrounding material when the water freezes.

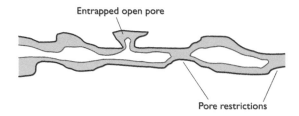

Figure 5   In real, complex capillary systems restrictions and entrapped pores create complex hydraulic pressures. Porosity becomes important for protection.

Real capillary systems are more complex again, with varying arrangements of wide, narrow, open and closed sections that trap moisture and create back pressures as the freezing front advances (Figure 5).

Porosity of lime mortar

Mature lime mortars, generally, have sufficient porosity to withstand most freezing conditions providing it is well designed with a well-graded sand. The addition of air-entraining admixture to a cement mortar is necessary because it doesn't have the natural porosity of a lime one. Addition of air entrainment to a well-designed lime mortar does not improve its frost resistance but does reduce its strength. Where a lime mortar is used in an exposed elevation with a risk of saturation then the addition of an integral waterproofing admixture will improve frost resistance. It acts to reduce the amount of water that can be held in the capillary system below that which, when frozen, expands beyond the pore volume. However, it will also reduce the permeability of the mortar to moisture vapour

Freeze–thaw cycling

The greatest danger from frost is when there are cycles of alternate freezing, thawing and re-freezing. As thawing takes place the mortar nearest the surface changes to water first. If a re-freezing occurs before the interior is fully thawed then complex hydraulic forces (Figure 6) can build up in trapped water between the two layers of ice, creating the frost damage typically seen with cracking parallel to the surface.

Even in the straightforward freezing and complete thawing cycle test results from the Foresight programme showed marked differences (Figure 7) for different mortar mixes after moist curing at 20°C for 91 days.[1]

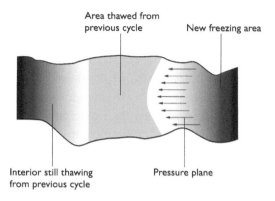

Figure 6 Complex effects of freeze–thaw cycling.

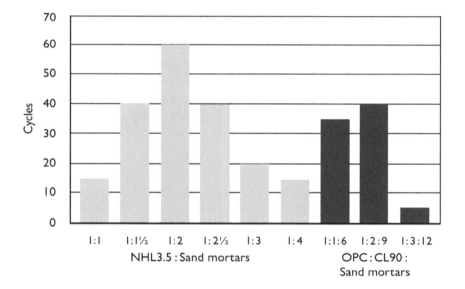

Figure 7 Resistance of mortars to freezing/thawing cycles.

Tests at CERAM

Tests commissioned by Lime Technology using the method adopted as the European test confirmed the above results. The method involves curing the test panels (ten rows each of three bricks wide bonded with 10 mm joints) for 28 days; immersing in water for 7 days; inserting into the chamber (rear face insulated); freezing for 100 minutes at –15°C; thawing for 20 minutes with warm air at 25°C; water spray for 2 minutes and drain for 2 minutes; repeat the cycle (ten cycles each 24 hours).

Three grades of mortar were tested: 'feebly' hydraulic (M 1); 'moderately' hydraulic (M 2.5); and 'eminently' hydraulic (M 5). The feebly hydraulic mortar showed signs of damage at 40 cycles; had lost 10 mm at 50 cycles; and lost 20 mm at 100 cycles. The moderately hydraulic mortar survived to 100 cycles without damage as did the eminently hydraulic mortar.

Moderately hydraulic and eminently hydraulic mortars were tested for the effect of curing period. After just seven days curing the moderately hydraulic mortar showed damage at 20 cycles and had lost 20 mm by 100 cycles; the eminently hydraulic mortar showed damage at 50 cycles and had lost 5–10 mm at 100 cycles. After fourteen days curing the moderately hydraulic showed damage at 50 cycles and lost 5 mm at 100 cycles, while the eminently hydraulic had superficial damage at 100 cycles. After 21 days curing the moderately hydraulic showed superficial damage at 50 cycles that did not deteriorate further at 100 cycles, while the eminently hydraulic showed no damage. The obvious conclusion is that strength class and curing time are critical factors in mortar durability.

Two types of mortar mixing systems were evaluated, a high shear mixer and a normal drum mixer. The type of mixer did not affect the freeze–thaw performance but the high shear mixing did result in 25% increase in strength.

Seasonal working practice

The key factors to achieving successful mortar work in extreme, particularly cold, conditions have been demonstrated. The first essential requirement is good planning and knowledge of the weather patterns for the area. The Meteorological Office can provide local historical records for specific UK postal code areas and these can be used to estimate when critical cold periods are likely. Past records for as much as five years might be necessary as they do vary significantly from year to year as Figure 8 demonstrates.

While there were no significant cold spells before the end of December in 2009, there were dangerously severe spells from the end of November in 2010. Taking the rate of gain in maturity previously demonstrated in Figure 1 and a rule of thumb that temperatures between 10 and 15°C extend maturity time by a factor of two and those between 5 and 10°C extend times by a factor of four, it can be seen from the 2010 data that full frost resistance according to CERAM tests would only be achieved by

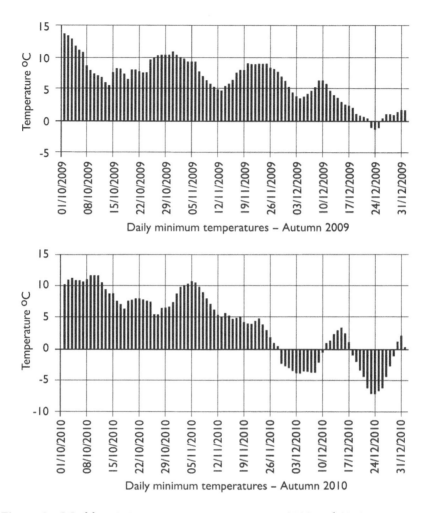

Figure 8   Weekly minimum autumn temperatures 2009 and 2010.

mortar placed early in October. Local microclimate variations must also be considered such as frost pockets and prevailing wind.

Nevertheless, additional precautions can enable work to go ahead in even the most severe conditions. A classic example of this was the restoration of garden walls at Hardwick Hall, undertaken by William Sapcote and Sons Limited and supervised by Stafford Holmes of Rodney Melville and Partners. Here the entire working and curing area of the wall was tented over and fitted with heaters to maintain temperature during the winter. Care was also taken to protect material storage and mixing areas.

WILLIAM SAPCOTE & SONS LTD
HARDWICK HALL – TEMPERATURE READINGS

LOCATION OF THERMOMETER: WALL 5

DATE	OVERNIGHT MINIMUM TEMPERATURE	TEMPERATURE @ 8.00AM	TEMPERATURE @ 1.30PM	TEMPERATURE @ 4.30PM	AVERAGE TEMPERATURE FOR DAY
22/2/01	INSIDE 7°C / OUTSIDE 4°C	9 °C / 7°C	MAX 13°C / MAX 10°C		
23/2/01	INSIDE 9°C / OUTSIDE 6°C	8°C / 8°C	MAX 14°C / MAX 12°C		
26/2/01	INSIDE 6°C / OUTSIDE -3	6°C / -1°C	MAX 9°C / MAX 0°C		
27/2/01	INSIDE 6°C / OUTSIDE -3°C	8°C / -1°C	MAX 10°C / MAX 0°C		
28/2/01	INSIDE 7°C / OUTSIDE -1°C	9 °C / 0°C	MAX 15°C / MAX 1°C		
1/3/01	INSIDE 6°C / OUTSIDE -5°C	8 °C / -1°C	MAX 12°C / MAX -1°C		
2/3/01	INSIDE 8°C / OUTSIDE -5°C	9°C / -5°C	MAX 8°C / MAX 0°C		
5/3/01	INSIDE 7°C / OUTSIDE -7°C	7°C / -3°C	MAX 12°C / MAX 8°C		

Figure 9 Good recording of site temperatures.

The insistence on good temperature measurement and records in the relevant internal and external areas was crucial (Figure 9).

Too often we see good intentions on these matters at the outset of repair contracts, but critical failures in securing protective sheeting, and in recording maximum and minimum temperatures in the critical work areas. This can be compounded by failing to monitor the exposed elevations during curing while attention is diverted elsewhere on a site. Simple precautions can very easily be the difference between success and damaging failure.

Similar questions arise regarding when it is safe to commence unprotected work at the start of the year. Figure 10 shows the variability in the incidence of late damaging frost, and snow in April is not unheard of in the UK.

The influence of curing conditions on mortar performance has been shown in Figure 2. Hot weather can occur at any time of the year but the most dangerous periods are during the summer when heat, low humidity and/or drying wind can cause work to dry out before hydraulic maturity can be achieved. Protection is again necessary to provide cooling shade, both for the work and for materials. Light spray to maintain work moisture should also be considered. Again accurate climatic records are essential to design and justify site actions.

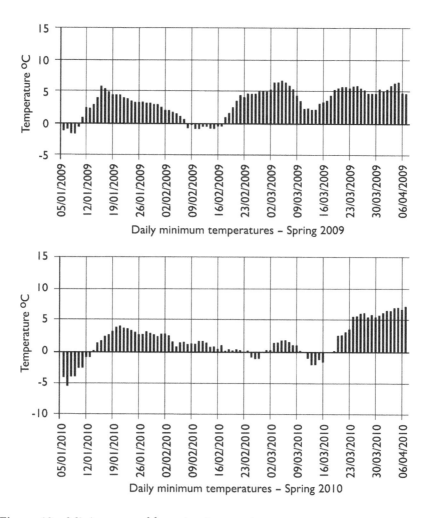

Figure 10   Minimum weekly spring temperatures.

THE CONCLUSIONS

While the climate in the UK is generally very favourable for working with limes, simple precautions including mix design, sand grading, porosity and curing conditions are essential to ensure successful outcomes in periods of particularly cold and hot weather.

Specification and work instructions should require:

- keeping of precise records of site and work climatic conditions;
- specific precautions to be taken against frost or heat;

- that no work should be permitted in cold weather when the temperature is 5 °C and falling;
- that work should be protected until cured when snow and frost are forecast;
- that work should be kept moist during placing and curing in defined high temperatures and dry conditions;
- that adjustment of the mortar specification (particularly binder hydraulicity) to offset a prolonged protection period is only permitted where the masonry design can accommodate it, and prior written approval should therefore be obtained;
- that the protection period should be extended in cold, wet conditions to offset the slower hydraulic chemical reaction;
- that exposure to high temperature is to be avoided as, although accelerating the hydration rate, it reduces the ultimate strength of hydraulic mortar.

This paper was first published in the *Journal of the Building Limes Forum*, Volume 16, 2010.

Note

1 Allen, G., Allen, J., Elton, N., Farey, M., Holmes, S., Livesey, P. and Radonjic, M., *Hydraulic Lime Mortar for Stone, Brick and Block Masonry*, Donhead Publishing, Shaftesbury, 2003.

Hot-Lime Mortars

A Current Perspective

Alan Forster

INTRODUCTION

Analytical and documentary evidence indicates that hot-lime mortars were used in traditional construction. These are defined as mortars manufactured by mixing quicklime and sand, rather than the current and more commonly adopted method of combining previously slaked lime with sand. Hot-lime mortars are again being used and are perceived to have advantages over cold manufactured mortars. Little is understood, however, regarding the physical and chemical performance of hot-lime mortars. This paper highlights current views on the subject and considers why hot-mixed mortars appear to offer better performance than alternative lime mortars. The potential for altered hydraulic properties, bond characteristics between the lime–aggregate interface, pore structure development and micro-structure performance are assessed, although it is stressed that this is only a preliminary investigation aimed at stimulating debate and further research.

Traditional mortar manufacture

Hot-lime mortars appear to have been commonly used in the construction of historic buildings, with evidence supporting this derived from available documents and from the results of an ever increasing quantity of mortar analyses.[1,2,3] Analysis of traditional mortars indeed indicates that they tend to be binder rich (sometimes in the region of 1:1 to 1:2.5 (binder to aggregate respectively) and have a quantity of whitish coloured, unmixed lumps of binder known as lime inclusions (lime inclusions may appear as unmixed dry powder, unmixed lime putty, reprecipitated lime or hot-lime inclusions; hot-lime inclusions may be distinguished by thin-section analysis and are generally fractured in nature when compared with the other forms). This view is shared by Gibbons, in work undertaken by the Scottish Lime Centre Trust, who believes that 'mortars were frequently made by the traditional method of slaking quicklime and sand together in one operation, often being used for building work immediately.'[4]

Hot-lime mortars are manufactured from a mix of quicklime (calcium oxide, CaO) and sand, giving an exothermic reaction, rather than the more common modern method of combining slaked lime (calcium hydroxide, $Ca(OH)_2$) in a dry hydrate (anhydrous) or putty form (aqueous) with sand. Hot-manufactured materials would either have been mixed and immediately used while the slaking process was live, or mixed and then left for a specified period depending upon the nature of the quicklime and the construction process being undertaken. However, it is important to realize that hot-lime mortars were not always used when hot and still slaking.

Today a small number of contractors use hot-lime mortars for certain applications for various reasons, principally because the mortars appear to be exceptionally durable, more safely used for winter working, have excellent workability characteristics, and are cheaper to manufacture (due to the low cost of the raw materials and the potential saving due to the doubling in volume that occurs when quicklime slakes into calcium hydroxide).

Hot-lime mortars, manufactured in several different ways, have historically been used in many construction situations, including brickwork, rubble masonry, wall cores and bridge construction. The manufacture of hot-lime mortars for use with brickwork has important differences, for example, from those for mortars used for the construction of wall cores. A longer manufacturing time is required with brickwork mortars to allow for the volume increase associated with slaking to cease prior to use. If this principle is not adhered to, 'jacking' and potential failure of the masonry structure might occur as the mortar joints expand. This phenomenon is not considered to be as problematic when constructing wall cores or mass masonry structures, as they have sufficient dead load to resist the expansive forces imposed upon them. Indeed, this may have been considered advantageous as the expansion in volume may have aided the filling of voids within the wall core and potentially reduced shrinkage during curing.

Traditional hot-lime mortars (hot-mixed mortars)

References can be found to the manufacture of hot-lime mortars, including British Standard Code of Practice 121, published as late as 1951.[5] However, the following quote, taken from Rivington's *Notes on Building Construction*, is unequivocal:[6]

> The proportion of ingredients in mortar is generally specified thus: '1 quicklime to 2 (or more) of sand', meaning that 1 measure of quicklime in lump is to be mixed with 2 measures (or more) of sand.

The quantities of quicklime to sand for a range of limes and cements are given in Table 1.

Table 1   Standard proportions of quicklime to sand.[7]

Binder type	Parts by measure [by weight]	
	Quicklime	Sand
Fat limes	1	3
Feebly hydraulic limes	1	2½
Hydraulic limes (such as Lias)	1	2
Roman cement	1	1 or 1½
Medina	1	2
Atkinson's	1	2
Portland	1	5
Scott's	1	4

The following extract from Rivington's *Notes on Building Construction* explains the basic process of slaking lime and sand together:[8]

> Preparation and mixing – The quicklime and sand having been procured, and their proportions decided, the preparation of the ingredients commences.

> Slaking – A convenient quantity of the quicklime is measured out on to a wooden or stone floor under cover, and water enough to slake it is sprinkled over it.[9] The heap of lime is then covered over with the exact quantity of sand required to be mixed with the mortar; this keeps in the heat and moisture, and renders the slaking more rapid and thorough.

> In a short time – varying according to the nature of the lime – it will be found thoroughly slaked to a dry powder.

> In nearly all limes, however, there will be found over-burnt refractory particles and these should be carefully removed by screening – especially

in the case of hydraulic limes; for if they get into the mortar and are used, they may slake at some future time, and by their expansion destroy the work.

Hydraulic limes should be left (after being wetted and covered up) for a period varying from twelve to forty-eight hours, according to the extent of the hydraulic properties they possess; the greater these are, the longer will they be in slaking. Care should be taken not to use too much water, as it absorbs the heat and checks the slaking process.

Jedrzejewska illustrates a variation on the hot-lime theme:[10]

Another method used in ancient times to improve the properties of lime mortars was to prepare a mixture of well slaked lime, sand and eventually, hydraulic ingredients, and to add to this, just before use, some proportions of freshly burnt quicklime. Such a 'hot lime' had to be used immediately.

Jedrzejewska continues by citing Lange: 'This method was used by the Romans. It was certainly known and used in later times.'[11]

The specification of Thomas Page for the New Bridge at Westminster states:[12]

Of lime – the lime used on the works shall be approved by the engineer before the quantity be ordered by the contractor. The lime shall be thoroughly burned, but quite free from core, and shall be used hot from the kiln, where practicable. In other cases, where it may remain a short time on the works, it shall be thoroughly protected from the air and moisture. If any lime core be found after the slaking of the lime, it will subject the contractor to the rejection of the whole of such lime, which he may have provided for the works.

Page's specification indicates a number of points, particularly that lime was often produced on or close to the building site and that the quicklime may have been used straight from the kiln.

Certain tentative factors may be drawn from the aforementioned information:

- quicklime was often slaked with sand for an initial period prior to full manufacture;
- partial slaking would have occurred after the 24- to 48-hour period, with a higher proportion of calcium oxide remaining unconverted when compared with modern highly refined calcium hydroxide;
- screening prior to final mixing/slaking appears to have been undertaken; this would have been essential for brickwork mortars;
- the quicklime adopted for construction may have been non-hydraulic or hydraulic depending upon its use;

- generally, traditional mortars were binder-rich, especially when compared to modern materials;
- quicklime may have been used straight from the kiln;
- characteristics of the mortar appear to be modified by hot-lime mixes and were favoured by many;
- the duration of slaking with aggregate depends upon the hydraulicity of the material/quicklime.

Modern forms of hot-lime mortars (hot-mixed mortars)

Today hot-lime mortars and gauged hot-lime (a binder consisting of a mix of non-hydraulic quicklime and hydraulic lime) mortars are currently being used by several experienced contractors in Scotland. There are a number of possible methods for the manufacture of hot-lime mortars. The first of these is to mix sharp (coarse) damp sand and non-hydraulic quicklime, bank it up in a heap, and leave to partially slake in situ for approximately 24 hours. The intensity of slaking and associated heat generation is directly related to the quantity of water contained within the sand. The second method is to combine damp sand and quicklime in a forced-action mixer, adding water little by little until slaking and workability have been achieved. Due to the highly refined nature of modern non-hydraulic quicklimes, slaking is achieved rapidly, especially if 6 mm chippings (kibbled lime) or powder are used instead of traditional lump lime (quicklime taken from a kiln in lump form is also known as shell lime). The addition of hydraulic lime may be made after initial slaking if a gauged hot-lime mortar is required.

The performance and manufacture of these mortars are little understood and only anecdotal evidence and empirical testing suggest their benefits. The following discussion seeks to highlight the perceived advantages of these materials and attempts, where possible, to explain some of the physical and chemical characteristics.

PERCEIVED ADVANTAGES OF HOT-LIME MORTARS

The benefits of heat generation

All previous references to the production of hot-lime mortars mention the combining of quicklime and sand and the leaving of them to slake for a specific period prior to use, depending on the nature of the quicklime. From this, it is not unrealistic to assume the following:

- generation of heat within the mortar during slaking may have a positive effect on the bond characteristics between the aggregate and lime;
- development of steam may create an altered pore structure within the mortar.

These apparent phenomena may also have the potential to alter the physical characteristics of a mortar, especially in hydraulic limes; this is considered below. The first of these actions, namely the lime–aggregate interface, may greatly affect the bond characteristics of a mortar and consequently its strength.

Typically, many commercially available non-hydraulic lime mortars (particularly those made from over-wet lime putties) are relatively weak and friable (but enjoy high flexural response). Although the putty is relatively sticky or 'fat', its relationship with certain aggregate types may, however, be poor when carbonated. Lewin states that, 'the poor mechanical strength of lime putty sand mixes occur due to the limited inter-particle fusion developing at calcite–calcite contacts and secondly, the small affinity of calcite crystallites for quartz surfaces.'[13] This opinion is supported by work undertaken by the author using environmental scanning electron micros-copy (ESEM). Figure 1 shows the surface of the aggregate with sporadic calcite crystallites bound to the surface on the right. In the upper left an agglomeration of calcite (carbonated non-hydraulic lime) can be seen. Located within the centre of the figure is a pore running from the upper right to the lower left of the image.

It is important to realize that, generally, the strength of any binding material may be correlated with the surface area of the crystalline and/ or amorphous matrix that is formed via hydration. The relatively low surface area of the 'chunky' calcite crystallites is therefore one of the most predominant factors influencing the strength of the mortars. This has been explained by Illston and Domone, who state that:[14]

> The strength of hardened cement paste (hcp) derives from Van Der Waals type forces between hydrate fibres [or products of hydration generally]. Although these forces are of relatively low magnitude, the integrated effect over the enormous surface area is considerable.

Those limes with a hydraulic component will have a higher relative surface area due to the nature of their crystallites than non-hydraulic counter-parts and will therefore have a greater strength. This can be illustrated by

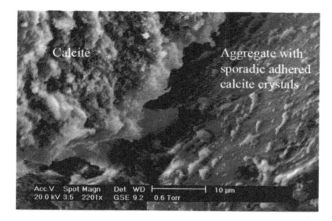

Figure 1 ESEM photomicrograph of non-hydraulic lime putty and coarse stuff mortar.

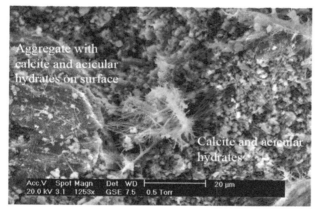

Figure 2 ESEM photomicrograph of hydraulic lime (St Astier NHL3.5) and coarse sand. Magnification is 50% of Figure 1.

looking at cement hydration in which 1 kg of anhydrous powder will yield micro-crystalline hydrates with approximately 200,000 m² surface area.

In comparison, the physical structure and aggregate–binder interface of a moderately hydraulic lime mortar (St Astier NHL3.5 and coarse sand, manufactured cold) as shown in Figure 2 is noticeably different. On the left side a piece of aggregate can be clearly seen. Relatively 'chunky' calcite crystallites can be seen sporadically adhered to the surface of the aggregate, correlating with Figure 1. Additionally, small rod-like (acicular) crystals may be seen and these are some of the hydraulic components responsible for the higher strength of hydraulic limes. These crystals clearly have a different shape to calcite and appear to be well adhered to the aggregate. On the right side of the image, the substrate is formed predominately of calcite interspersed with acicular hydrates throughout the surface and body of the material

The interaction at the cement–aggregate interface continues to be discussed by researchers. Odler states:[15]

The structure of this 'transition zone' has been extensively studied. Most investigators reported the presence of a thin continuous film, consisting of calcium hydroxide, adhering to the aggregate surface. It was also suggested that a layer of rod shaped C-S-H gel particles covers the calcium hydroxide layer, producing an assemblage called the 'duplex film' with an overall thickness of about 1–2 mm.

He continues to state that:[16]

> … in contrast, investigations performed by Scrivener and co-workers gave no indication of the presence of a duplex film or of a continuous layer of calcium hydroxide at the rock surface. Instead it was reported that the phase which is in immediate contact with the aggregate surface most often is C-S-H. [Cement chemists adopt a slightly different chemical notation, such that C refers to calcium rather than carbon.]

It is currently believed that the strength characteristics of a non-hydraulic lime mortar may be improved by hot manufacture. The process is understood to enhance the mechanical bond strength via scarification of the aggregate. This process is explained by Gibbons, who states that 'during the mixing process the action of the hot caustic lime on silica sand can potentially etch the surface of otherwise unreactive silica grains.'[17]

If the principles of etched aggregate surfaces are assumed to be realistic in hot-lime mortars, the bond characteristics and ultimate performance should be enhanced whether the resulting mortar is used hot or cold.

Additionally, the bond characteristics between the aggregate and the lime may be enhanced by a secondary phenomenon (though not confirmed through research): namely the dehydration of aggregate surfaces. This is a process by which the water surrounding and contained within the sand particle is dehydrated during the heat generation of slaking. Gibbons believes that, 'the initial slaking process has the effect of heating and drying the sand grains, which then form a good bond with the lime paste as water is added.'[18] This concurs with work by Jedrzejewska.[19] In addition, if water is lost then porous aggregates (such as crushed sandstones and limestones) may theoretically have a greater affinity for water on initial wetting, and so readily adsorb it into the aggregate pore structure. When the water is drawn into the aggregate, it is not unrealistic to expect that fine particles of calcium hydroxide will be transported into the body of the sand, thereby creating a mechanical bond that may lead to longer-term strength in the mortar as the calcite–aggregate bond develops. This interpretation cannot be verified without further research. In addition, such an explanation may be an over-simplification of the physiochemical reactions taking place. The wettability characteristics of the aggregate, which are a function of the adsorption/surface characteristics of the material, will, for instance, have a bearing upon the capillarity of any porous material. In most basic terms are the surfaces hydrophilic or hydrophobic in nature?

The conferment of hydraulic reactivity and effects of temperature on the hydration of limes and cements

The use of quicklime on a specific project would have required the material to be obtained close to the site as transporting from a distant source would increase the cost of the works. Indigenous Scottish limestones primarily produce hydraulic limes as they contain silica, aluminium, iron and other minerals that combine with the lime during firing to form reactive hydraulic compounds (calcium silicates, calcium aluminates, alumino-silicates, and alumino-ferrites). The type and reactivity of these hydration products are determined and influenced by many factors, but with the composition of the stone and the temperature of firing within the kiln being the primary influences.

The nature of these hydration products may be estimated by the use of phase diagrams, generated from the mineralogical composition of the stone twinned with the firing temperature of the kiln. The formation of dicalcium silicate (C_2S) (also known as belite) may be attained from relatively low firing temperatures (850°C), twinned with the correct mineral composition. This is the principal hydraulically reactive component found in hydraulic limes. Dicalcium silicate is a hydraulically reactive salt that is slow to hydrate in the presence of water and hence hydraulic limes tend to have a slower initial strength development than ordinary Portland cement (OPC). However, the same mineral composition fired at a higher temperature, such as 1,350°C, will lead to the formation of tricalcium silicate (C_3S) (or alite).

This is the main reacting compound in OPC, but in theory it should not be evident in hydraulic limes due to their lower firing temperature. In practice, due to variable temperatures within a traditionally fired kiln, small quantities of tricalcium silicate can be found in historic hydraulic lime mortars. Alite is fast-reacting when compared with belite, and may form a dense impenetrable matrix of amorphous and crystalline hydrates, which explains why OPC is a denser and stronger material with less flexural response.

It can therefore be seen that the nature of the hydration products will be reflected in the performance of the binder. When an anhydrous compound (such as C_2S) has been hydrated, it will react with water and form calcium silicate hydrates (C-S-H). However, if the composition of the limestone includes aluminium, calcium aluminate hydrates (C-A-H) will also be formed.

With regard to the conferment of hydraulicity in mortars via hot-lime production methods, the alteration of hydraulic properties has been discussed in work by Jedrzejewska: 'it seems that far more important than the ingredients are the methods used to prepare the mixture and, especially, the lime.'[20] She emphasizes that 'a considerable amount of heat is evolved in the reaction between calcium oxide (quicklime) and water.'

Jedrzejewska continues on the theme, citing Lange:[21]

> The temperature may rise to 270–300°C, and even over 450°C in special conditions [conditions not specified]. This heat is practically lost when quicklime is slaked in open boxes. But if slaked in a closed space, the amount of heat evolved and the overheated steam, would exert a very strong chemical action on any siliceous materials present, bringing them, at least partly, into solution. Lime, slaked by this method, would have quite different properties to lime from the same raw materials slaked by the open-box method. This possibility is strongly reflected in a series of German patents of the nineteenth century at an early period in the use of Portland cement. Prescriptions are given for various lime mortars and cements prepared from quicklime slaked in closed containers in the presence of different siliceous materials such as powdered low-burnt brick, certain kinds of sand, ashes of brown coal, powdered pumice stone, trass, iron slag, diatomaceous earth, clay, Portland cement, and asbestos, etc. All this led to mortars of greatly improved qualities due to the presence of active hydraulic components.

Jedrzejewska continues: 'There are certain indications that such a method could have been used for the preparation of ancient mortars, much earlier than [the] nineteenth century.' Citing Berger,[22] she gives a description of a method of slaking quicklime in heaps well covered with sand, and only small amounts of water. According to Berger:[23]

> The method is probably based on much older traditions. The siliceous materials could be present in quicklime naturally from clay and other impurities, or they could be added intentionally before slaking. This could perhaps be another cause for some extraordinary properties of ancient mortars, so mysteriously impossible to reproduce, even when, apparently, the same ingredients were used. Experiments carried out by the writer, with lime mortar very similar to the kind of cold mortars heretofore impossible to reproduce …

It is unlikely that the addition of clay minerals to a hot-lime mortar mix, as discussed by Berger, would have conferred hydraulicity in the material.

The heat generation in the slaking process was quoted by Jedrzejewska as being between 270 and 300°C with an upper limit of 450°C. This does not correlate with our understanding of the conferment of hydraulicity in clays. This is best illustrated by looking at the dehydration characteristics of clay minerals, although it must be emphasized that they all have different properties (e.g. smectite will dehydrate differently from illite).

Clay consists of many platelets with water trapped in between them and this water is known as inter-laminar water. However, water is found within the clay mineral itself and this is known as its constitutional water. The amount of inter-laminar water changes with exposure to wetting and drying cycles. When the plates are subject to wetting, they swell, and conversely when they dry the clay desiccates and shrinks. This explains why clay soils heave and cause damage to buildings. Changes occur when clay is heated, some being reversible and others irreversible, depending upon the temperature. Generally, the conferment of reactivity in a clay mineral occurs when a sufficiently high temperature has been attained, and consequently the silicate has been converted from an insoluble material into a soluble material. Table 2 indicates the changes that occur in kaolinite at certain temperatures.

Table 2   Effects of heat on clay minerals (kaolinite).[24] These temperatures may vary from clay mineral to clay mineral depending on particle size and composition.

Temperature (°C)	Effect of dehydration on clay minerals	Reversible/ Irreversible
110	Some adsorbed water (surface and interlayer) is lost. This is common to all clay minerals.	Reversible
400	Loss of all surface and interlayer water.	Reversible
400–525	Loss of nearly all constitutional water (OH).	Reversible
650–800	Loss of nearly all water molecules and (OH) ions are driven off. Formation of meta-kaolinite and meta-dickite.	Reversible
800+	Layered structures are further disrupted and cannot be reconstituted by rehydration, finally decomposing into mullite and cristobalite.	Irreversible

It is therefore clear that at temperatures below 400°C no loss of constitutional water in silicate minerals occurs and therefore no hydraulicity should ensue. However, at temperatures above this, the conferment of hydraulicity may be achieved.

This view appears to be reinforced in work by Taylor, who emphasizes:[25]

> The behaviour of clay minerals on heating depends on their structure, composition, crystal size and degree of crystallinity. In general, any inter-layer or adsorbed water is lost at 100–250°C; dehydroxylation begins at 300–400°C and is rapid by 500–600°C … Above about 900°C, all begin to give new, crystalline phases.

These become reactive due to disorganization in the crystal structure, and could potentially contribute to hydraulic properties in the mortar.

In addition, in consideration of Berger's comment relating to clay impurities within the limestone deposit, the resultant quicklime would have been hydraulic in nature with hydraulically reactive compounds such as calcium silicate and calcium aluminate produced within the firing process.

The enhanced performance or initial conferment of hydraulic properties upon siliceous components or clay obviously requires further investigation; mild reactivity cannot, however, be ruled out.

Effects of temperature upon the hydration of limes and cements

It is understood that the temperature at which cement is hydrated is positively related to an increase in the rate of the hydraulic reaction. This is one of the main reasons why pre-cast pre-tensioned concrete members are steam cured. It can be expected that the hydration of hydraulic components in hydraulic limes should also follow this principle, as it is believed that the mechanisms of set are similar in terms of C-S-H formation.

Upon initial contact with water the conversion of quicklime into calcium hydroxide, $Ca(OH)_2$, occurs with the generation of heat. If a minimum of water is adopted for the slaking process, the reaction of the hydraulic components will not occur. However, with the addition of extra water, hydrolysation of $Ca(OH)_2$ occurs and hydration (primarily a dissolution and precipitation reaction that occurs when water comes in contact with the hydraulic compounds) of aluminate, silicate and aluminosilicate phases begins. These ultimately yield hydrates, both amorphous and crystalline in nature. It is well understood that the solubility of hydraulically reactive components is increased when temperatures rise and consequently the hydration rate may speed up.

In a historic mortar the degree to which the quicklime would have converted to $Ca(OH)_2$ would depend upon the quantity of water and duration of slaking. The quantity of unslaked quicklime present, even after a longer period of slaking, would be higher than that found in modern commercially produced dry hydrates [anhydrous $Ca(OH)_2$]. The greater the proportion of quicklime, the higher the temperature of the mortar being finally manufactured (the conversion rate of quicklime is also dependent on the extent of over- or under-burning).

This may have an effect upon the performance of the mortar, and is noted in work by Odler relating to the hydration of alite (C_3S) stating, 'the rate of C_3S hydration is also temperature dependent and increases with temperature.'[26] This is an extremely well researched area and the rate of reaction has developed complex equations to model the hydration kinetics of C_3S (i.e. Avrami's equation for acceleration kinetics and Jander's equation for deceleration kinetics). The reaction has been little researched in C_2S (the main cementing compound in hydraulic lime), although it is well understood that C_2S generally hydrates at a much slower rate than C_3S.

In addition, the mechanisms of hydration are believed to be very similar to each other, this being reinforced in work by Jawed, Skalny and Young, who indicate that, 'it can be concluded that C_2S reacts mechanistically in a way similar to C_3S.'[27] It is also interesting to look at work by Odler covering Portland cement hydration at elevated temperatures (hydration at 0–100°C). Odler indicates that:[28]

> The rate of hydration of Portland cement increases with temperature, especially at lower degrees of hydration, whereas this effect becomes less pronounced as the hydration progresses. Among the individual clinker mineral, only the hydration of belite (C_2S) was found to be accelerated significantly even months after mixing with water.

Considering that C_2S is the main cementing compound in hydraulic lime, it is not too unrealistic to expect some degree of altered activity in hydration of the materials.

Conversely, the solubility of $Ca(OH)_2$ decreases with an increase in temperature, the result being the formation of rather speedily developed portlandite crystals that may occur in a wide range of sizes, theoretically engulfing the C-S-H, C-A-H and/or calcium aluminosilicate hydrate [C-A-S-(H)] that are concurrently forming at an increased rate.

The use of hot materials may have an effect upon the rate of hydration reaction in the mortar; to what degree will require further research.

These reactions are only one part of an extremely complex area of study. The rate at which hydration of any hydraulically reactive compound can occur is greatly affected by the concentration of calcium dissolved in the liquid. Once the solution becomes saturated with calcium, the rate of the hydration reaction temporarily ceases (known as the dormant period) until the saturated condition reduces in concentration via the growth of calcium hydroxide (portlandite) in the liquid phase and/or via the formation of hydraulic compounds, such as C-S-H, that consume a sufficient proportion of the calcium.

This is reinforced by Jawed, Skalny and Young, believing that nucleation and growth of $Ca(OH)_2$ in C_2S is not necessarily as important as in C_3S.[29] They do, however, indicate that it is of utmost importance to have a 'calcium sink' to allow continued hydration to occur.

Hence, with regard to the solubility of $Ca(OH)_2$ in water, this reduces with an increase in temperature. This may be clearly seen in results by Boynton, in which he indicates that, 'the magnitude of solubility of $Ca(OH)_2$ is 1.330 g CaO/l of saturated solution of 10°C in distilled water, and at 0°C solubility increases to 1.4 g CaO/l.'[30] As opposed to limestone (calcium carbonate), the solubility of lime hydrate is inversely proportional to temperature and therefore decreases as temperature rises (Table 3).

Table 3   Solubility of lime expressed as CaO or $Ca(OH)_2$ at different temperatures (g/100 g saturated solution).[31]

Temperature (°C)	CaO	Ca(OH)₂
0	0.140	0.185
10	0.133	0.176
20	0.125	0.165
30	0.116	0.153
40	0.106	0.140
50	0.097	0.128
60	0.088	0.116
70	0.079	0.104
80	0.070	0.092
90	0.061	0.081
100	0.054	0.071

The ramifications of this phenomenon could possibly be twofold: first, to act as a calcium sink and thereby initiate rapid hydration of the hydraulic component in the lime/cement, and, second, to generate portlandite [$Ca(OH)_2$] crystallites over a wider order of magnitude (crystallite sizes may range from a nanometer through to several millimetres) than mortars made via cold processes.

Activation of mildly pozzolanic aggregates

It has been suggested that the heat associated with hot-lime mortar manufacture may lead to a reaction between one of the aggregate components (such as silica, alumina or iron) and the lime, i.e. the aggregate becomes mildly pozzolanic. A pozzolanic material differs from a cementitious material in that it will not set on its own. This is due to the absence of calcium oxide, which must be supplied to generate a hydraulic reaction. Hydraulic limes and cements have CaO present from their manufacture and therefore do not require any additional components to be added for reactivity to occur (when mixed with water). Gibbons explains that hot mixes are 'thought to improve the bond between lime and silica and perhaps to create some mild pozzolanic activity in suitable sands.'[32] This concept is reinforced in work by Jedrzejewska.[33] This appears unlikely in siliceous aggregates due to the relatively low slaking temperature previously mentioned, a view reinforced by Taylor, who indicates that, 'quartz undergoes a minor, rapid and reversible phase transition to a-quartz at 573°C. It is unstable relative to tridymite at 867–1470°C, and to cristobalite above 1470°C.'[34] Although this appears to question the increased conferment of hydraulic properties within hot limes, more research is required to confirm this. In addition, it is likely that those aggregates that are, by nature, mildly pozzolanic may favour this hot environment for rapid hydraulic component development, namely C-S-H.

Altered hydration characteristics in quicklime: ß*C$_2$S modification via quenching

As discussed above, the hydraulicity and reactivity of a binder may be affected by the composition and temperature at which it is fired. Generally speaking, cement reacts quickly and is relatively fast setting as it contains a high proportion of alite (C$_3$S) that hydrates quickly into C-S-H. Conversely, hydraulic limes hydrate slowly due to the predominance of belite (C$_2$S), which forms C-S-H. It is understood that approximately 90% of C$_3$S is hydrated before 28 days, while only 10% of C$_2$S has reacted after the same period.[35] This phenomenon is true with regular manufactured limes and cements, but research by Lawrence concerning active belite cements (namely those that have C$_2$S as the main hydraulic compound) has led to an interesting discovery.[36] When the oxide (calcium oxide + calcium silicate, aluminates and other hydraulically reactive compounds) is taken from a kiln, the way in which it is cooled may play a significant role in the hydration of the material. Lawrence indicates that 'the speed of hydration of ß-C_2S can be increased by imparting a stressed condition through thermal shock, or through the formation of small crystallites',[37] and emphasizes that this is achieved by altering the cooling rate of the material.

Importantly, if the oxide is rapidly cooled it may lead to the formation of a new form of ß-C_2S known as ß*C_2S. This is believed to exhibit rapid hydration characteristics and therefore faster set and strength development. This begs the question that if a hot form of quicklime was removed from a kiln and used almost immediately, could it have been cooled at a sufficient rate to form a new polymorph, ß*C_2S? Also, did traditional mortar manufacturers serendipitously manufacture fast-reacting forms of hydraulic lime?

The mechanisms are extremely complex and may not directly relate to hydraulic limes; however, it may be of some merit to study this process to assess the effects in this context.

Pore-structure development and pore-size distribution

When non-hydraulic lime putties are matured, the portlandite crystals [$Ca(OH)_2$] tend to decrease in size. This is ideal for materials such as lime plasters, as it ensures a uniform and consistent binder, free from unslaked particles and with excellent workability. When a hot-lime mortar is made or slaked lime is used without maturing the material, it may be reasonable to assume that the portlandite crystals that form do not attain the fineness that slaked and matured lime putties achieve. On the contrary, they may actually have a particle size over a wide order of magnitude. It is not therefore too unrealistic to expect the early-stage development of a hydraulic or non-hydraulic hot lime to have a pore structure that is coarser in comparison with a cold-manufactured coarse stuff. In addition, during carbonation of the material, it is expected that, although changes occur in the transformation of the calcium hydroxide into calcite, they should follow the skeletal macro-pore system established by the freshly applied lime mortar.

Further, although little researched, the use of lime mortars, if used in a hot state, may lead to the alteration of the pore structure due to forces generated by heat and steam. This is a view held by Gibbons, who believes that, 'the presence of steam, generated by the slaking process also has the effect of entraining air in the mortar and appears to improve the pore structure.'[38] If steam was generated within the mortar, this might lead to the development of a large interconnected pore structure contributing to its durability.

The possibility of the increased durability of mortars made and used hot can only be surmised at present. However, when thin-section analysis of hot- and cold-manufactured mortars is undertaken, it becomes apparent that differences in the pore structure do exist. It is clear from Plate 21, which shows a non-hydraulic lime mortar manufactured using lime putty, that the macro-pore structure is both interconnected and segmented in nature, with a pore-size distribution over a wide order of magnitude (those regions indicated). In Plate 22, showing a feebly hydraulic hot-lime mortar, the pore structure appears relatively cavernous and has a high degree of interconnectivity. Although some of the pores appear to be segmented it is important to remember that the images show a two-dimensional view and in reality the degree of interconnectivity would generally be greater. A large interconnected system may lead to greater levels of durability, due to the relatively cavernous pores. These may have the potential to accommodate freeze–thaw cycles and moisture transfer.

If a pore system were established with the aforementioned qualities, twinned with other hypothetical mechanical properties, it may go some way to explaining the perceived high resistance of these hot-lime mortars to frost damage. Additionally, if a macro-pore structure of this nature were developed, it may aid in the ability of the mortar to transport moisture more readily due to the higher degree of interconnectivity.

CONCLUSIONS

It is evident that differences exist in the manufacture and use of hot-lime mortars when compared to cold-manufactured materials. The degree to which altered characteristics occur is difficult to assess due to the lack of relevant data and only anecdotal evidence to support enhanced durability.

The physical properties of hot limes, especially the pore structure and bonding characteristics, appear to be key factors in their performance. Although most evidence suggests that hot limes manufactured in a closed system appear to be most successful, other methods of manufacture cannot be overlooked and it is important that more research is undertaken in this area. It is, however, important to realize that in Scotland the quick-lime adopted for hot-lime manufacture would in the past have been to greater or lesser degrees hydraulic in nature and therefore the macro- and micro-pore structure would have been of greater complexity.

For plates referred to in this paper, see the colour section following page 132.

This paper was first published in the *Journal of Architectural Conservation*, Volume 10, No, 3, November 2004.

Notes

1 Gibbons, P., *Preparation and Use of Gauged Hot Lime Mixes*, unpublished paper, 2000.
2 British Standard Institution (BSI), *Code of Practice 121.201, Masonry, Walls – Ashlared with Natural Stone or with Cast Stone*, British Standards Institution, London, 1951.
3 Smith, P., *Notes on Building Construction: Part 3 – Materials*, Rivingtons, London, 1875, reprinted by Donhead Publishing Ltd, Shaftesbury, 2004.
4 Gibbons, *op. cit.*, 2000, p. 1.
5 British Standards Institution, *op. cit.*, 1951.
6 Smith, *op. cit.*, 1875/2004, p. 206.
7 Ibid., p. 207.
8 Ibid., pp. 207–8.
9 This is clearly not advisable as slaking lime is an exothermic reaction and may burn through the floor.
10 Jedrzejewska, H., 'New Methods in Investigation of Ancient Mortars', *Archaeological Chemistry*, papers from the Symposium on Archaeological Chemistry, American Chemical Society, Washington D.C., 1967, pp. 156–7.
11 Ibid., p. 157.
12 Holmes, S., 'Hot lime in a cold climate', *Journal of the Building Limes Forum*, Vol. 1, No. 2, 1993, p. 27.
13 Lewin, S. Z., 'X-Ray diffraction and scanning electron microscope analysis of conventional mortars', *Mortars, Cements and Grouts Used in the Conservation of Historic Buildings*, ICCROM, Rome, proceedings of a conference held 3–6 November 1981, p. 106.
14 Illston, J. M. and Domone, P. L. J., *Construction Materials: their Nature and Behaviour*, 3rd edition, E. & F. N. Spon, London, 2001, p. 104.
15 Hewlett, P. C. and Lea, F. (eds.), *Lea's Chemistry of Cement and Concrete*, 4th edition, Arnold, London, 1998, p. 284.
16 Ibid.
17 Gibbons, P., *Technical Advice Note 1: Preparation and Use of Lime Mortar*, revised edition, Historic Scotland, Edinburgh, 2003, p. 28.
18 Ibid.
19 Jedrzejewska, *op. cit.*, 1967, p. 157.
20 Ibid., p. 156.
21 Ibid., p. 156.
22 Ibid., pp. 156–7.
23 Ibid., pp. 156–7.
24 Deer, W. A., Howie, R. A. and Zussman, J., *An Introduction to the Rock Forming Minerals*, Longman Publications, London, 1989, p. 257.

25 Taylor, H. F. W., *Cement Chemistry*, Academic Press, London, 1990, p. 73.
26 Hewlett, *op. cit.*, 1998, p. 245.
27 Barnes. P., *Structure and Performance of Cements*, Applied Science Publishers, London, 1983, p. 266.
28 Hewlett, *op. cit.*, 1998, p. 288.
29 Barnes, *op. cit.*, 1983, p. 267.
30 Boynton, R. S., *Chemistry and Technology of Lime and Limestone*, 2nd edition, John Wiley & Son, New York, 1980, p. 203.
31 Ibid., p. 204.
32 Gibbons, *op. cit.*, 2003, p. 28.
33 Jedrzejewska, *op. cit.*, 1967, p. 156.
34 Taylor, *op. cit.*, 1990, p. 73.
35 Neville, A. M. and Brooks, J. J., *Concrete Technology*, Longman Publications, Harlow, 1993, p. 15.
36 Hewlett, *op. cit.*, 1998, p. 442.
37 Ibid.
38 Gibbons, *op. cit.*, 2003, p. 28.

Hot Lime Mortar in Conservation

Repair and Replastering of the Façades of Läckö Castle

Ewa Sandström Malinowski
and Torben Seir Hansen

THE LÄCKÖ PROJECT AND HOT LIME

Läckö Castle is situated on a rocky peninsula on the southern shore of Lake Väner, the largest lake in Sweden. The castle is a large, mainly seventeenth-century baroque structure, built on medieval foundations with walls of stone and brick covering a lime-concrete core. The castle was enlarged to its current size by Magnus Gabriel De la Gardie (1622–86), the King's Chancellor, and one of the greatest landowners in Sweden.[1] It is an important national monument and tourist attraction, owned and managed by the National Property Board.

The last major restoration of the façades of Läckö Castle was carried out in the 1960s. Over a period of ten years, the exterior was totally replastered with a hard, cementitious mortar, partially reinforced with chicken wire. The masonry walls were grouted at the same time with lime-cement mortar. Signs of decay, plaster detachment and algae started appearing on the façades more than a decade ago. Today the decay is more serious: brick damaged by frost, rotting wooden joists, moisture and salt in the interiors; all this clearly stems from the work undertaken during the 1960s.

A research project, 'Historic Mortars at Läckö Castle', run jointly by the National Property Board and the Department of Conservation at the

Figure 1 Läckö Castle, the southern façade, entrance to the Castle shown before the façade facing the Castle's southern courtyard was replastered in 2009.

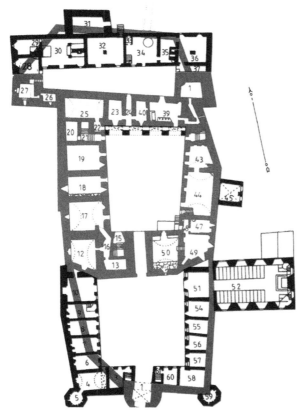

Figure 2 Ground plan.
Grey: Middle Ages.
Black: seventeenth century (Courtesy of Ragnar Sigsjö).

Figure 3   Damaged wooden beam (one of five beams) supporting the stone wall above window frame and niche.

Figure 4   Rubble core wall (stone and mortar) revealed after the removal of the damaged beams.

University of Gothenburg, started in 2002. This research developed into a full-scale repair and conservation project. In response to the significantly progressing damage, the decision was taken by the National Property Board to gradually replaster all the façades of the building.

The objective of the Läckö project has been to:

- develop a repair mortar that is historically and technically compatible with that in the original structure and which is produced using local materials;
- recreate 'original' methods of production and application and adjust these to modern conservation practice.

The initial phases of the project included experiments with local lime burning and production, using limestone from the nearby Kinnekulle Hill. Kinnekulle lime has feebly hydraulic properties, and has traditionally been considered a lean lime. It is no longer available commercially.

Sampling and testing original mortars

Sampling, testing and documentation of the surviving original, historic mortars was initiated.

Original mortars could be found in the masonry bedding, the wall cores and on a few formerly exterior surfaces that are now built in. These were found to be extremely lime-rich with a lime–sand ratio ranging from 6:1 to 1:1 by volume. The richest samples were from De la Gardie's era, the seventeenth century. The medium-rich samples were found in the medieval construction. The leanest examples came from the eighteenth-century repairs. Historic samples were examined wherever they were exposed by removal of the cementitous plaster.[2]

Experiments in the reproduction of lime-rich, historic mortars with lime–sand ratios 1:1, 2:1 and 4:1 by volume have been conducted on limited trial surfaces in recent years. The intention was to try to understand the historic crafts and methods, and to learn from the experience. The use of these lime-rich mortars on larger surfaces has proved problematic, principally because of shrinkage problems.

Experiments with replacement mortar formulation

In order to find a suitable compromise between historic and modern plastering practices, the first experiments with replacement mortar formulation and plastering were carried out using conventionally slaked lime putty produced from the local Kinnekulle lime. A series of mortars with different binder–aggregate ratios, varying within the limits 1 : 1 and 1 : 2.5 (by volume), was tested and evaluated. However, mortars richer than 1 : 1 were excluded as impracticable. Mortar samples based on commercially available lime, as well as different hydraulic lime mortars, were also tested as reference points.[3]

Experiments with hot-mixed lime commenced in 2002 as it was felt that this process could well provide the key to understanding how to achieve mortars with high binder contents similar to the originals. The incidence of these experiments gradually increased and the first full-scale trial was carried out in the summer of 2007. A total of 350 m² hard, cementitious render from the 1960s was replaced by locally produced lime mortar: 300 m² by hot lime mortar plaster and 50 m² by mortar made from conventionally wet-slaked lime putty and sand as a reference. Hot-mixed and wet-slaked techniques continued in parallel throughout the whole test series in order to enable thorough comparison between them.

The most recent phase of replastering on a large scale was completed in 2009 on the main, southern façade. This whole façade, measuring 360 m² in total, has been replastered with hot lime mortar, but has not yet been evaluated.

WHY HOT LIME?

The experiments with hot lime started at Läckö as an attempt to achieve 'fat', binder-rich mortar which could easily be applied to the façades. According to views expressed in current publications hot-mixed lime is thought to: [4,5]

- improve the bond between lime and sand;
- create a mild pozzolanic activity in sands;
- improve the pore structure in mortars;
- improve working properties;
- give durable mortars.

Slaking quicklime under cover of sand is described by Philibert Delorme as early as the sixteenth century.[6,7] A Swedish architecture manual from the nineteenth century describes slaking quicklime with sand as an appropriate method for producing mortar from lean or feebly hydraulic lime.[8] In both sources, the heat developed during slaking is considered crucial for the quality of the mix, although the resulting lime–sand mix in both the

Delorme and Swedish case is applied cold. It is possible to apply mortars while still hot from the slaking, particularly if the lime mortar and lime concrete is to be used in cold climatic conditions.[9]

It is quite probable that the rubble-core walls at Läckö Castle were constructed with the hot-mix technique as the expansion of lime during the slaking process assists compaction when used in a confined space. However, very little is known about the historic, exterior plastering techniques used as there are only very limited traces left. It is known that lime burning had been undertaken near the building site; however, no traces of lime pits have been found. This suggests that matured putty was not used in appreciable quantities as a binder, and that hot lime might therefore have been used as an alternative.

It was not the original intention that hot-mixes should be developed to prove their superiority as a technique. However, the experiments with hot-mixes, used in cold state, gave initially positive results and the material satisfied the craftsmen. An important argument in favour of hot-mixes, which has been put forward historically and currently, is that they do not require lime to be burned and stored for a year in advance of the production of hot lime mortar. However, this is necessary when putty mortar is produced.

In the following text quicklime slaked together with sand is referred to as 'hot lime'. Mortar produced with this lime, although applied cold, is referred to for convenience as 'hot mortar'. Lime slaked through conventional slaking to lime putty is referred to as 'wet lime'. Mortar produced with this lime is referred to as 'putty mortar' or 'wet mortar'. The resulting plaster on the façades, made with hot lime as binder, is referred as 'hot mortar plaster'. Plaster made with wet lime binder is referred to as 'wet mortar plaster'.

PRODUCTION OF HOT AND WET MORTARS

Raw materials

The sources of information about the materials that were originally used at Läckö are the building itself and written sources such as the correspondence between De la Gardie and his contractors, Swedish and German master builders. According to the correspondence, the lime was burnt a short distance from the castle, possibly together with bricks in the same kilns. However, no traces of lime pits have been found near the castle.

Today, the limestone is quarried in Kakeled, on the western slopes of Kinnekulle Hill, in the vicinity of the original seventeenth-century quarry and in the same geological stratum. The stone originates from relatively pure limestone beds interlayered in alum shales (Upper Cambrian to Lower Ordovician). These beds consist of dark grey limestone concretions with a high content of calcium carbonate (about 88–90% by weight). Apart from calcium carbonate, the stone contains impurities of clay minerals, pyrite, apatite, quartz and graphite as well as small amounts of rare trace elements.[10] The hydraulic properties of the lime are mainly related to the amount of quartz and clay minerals in the limestone. According to thin-section microscopy, the mineralogy of the binder is similar in both newly produced and historic mortars.

The feebly hydraulic lime produced at Kinnekulle for hundreds of years had a good reputation as a binder for both plasters and stuccoes. It was exported and sold in the form of quicklime. However, according to local oral tradition, wet-slaked lime had always been used in the past for building purposes.

Fine-grained siliceous sand from the lake (grains mainly up to 0.5 mm) was originally used as the aggregate in the historic mortars, but the sand-banks on the lakeshore are now protected. Today the natural sand aggregate is brought from a quarry 20 km away from Läckö. It consists of quartz, feldspar and fragments of granite/gneiss; the grain form is predominantly edge rounded, and the sand grains can measure up to 6 mm. The mineral composition and grain form of the sand now used is comparable to the historically used aggregate.

Lime burning

The experiments in burning lime (technically expressed as calcination) were carried out in a small, experimental lime kiln built near the castle. When larger quantities of lime were required (in 2004 and 2005), an old, traditional tunnel kiln, normally used for brick production, was used. Both kilns were fired with wood. Today an experimental kiln in Mariestad, belonging to the Department of Conservation, University of Gothenburg, is used. This is fuelled by wood, mostly fir and pine. It is possible that charcoal and hardwoods, such as oak, might have been used originally, raising the temperature considerably.[11] Thus the lime produced today might be burnt at a lower temperature than in the past; 850–950°C in the central part of the kiln.[12] Microscopic analysis of thin sections has partially confirmed these assumptions: the historic mortars contain a

Figure 5   Lime burning in experimental kiln, Department of Conservation, University of Gothenburg, Mariestad.
Figure 6   (top right) The kiln is loaded from above.
Figure 7   (bottom right) Wood is used as fuel.

greater variety of particles with hydraulic properties compared to the ones that are produced at Läckö today.[13] Grains with traces of ferrite phase, (C_4AF), and calcium silicate phases (C_2S and occasionally C_3S) are more frequent in the historic mortars.[14,15]

The lower burning temperature in the current process gives 'softer' burning. Unburnt grains of limestone are found more frequently in the mortars produced today (on average 3.5% of binder volume) than in the original, historic mortars (less than 1%).

A few sintered lime particles with retarded slaking that disrupted the plaster surfaces of the early experiments (producing popping and pitting) indicated high temperatures near the hearth in the burning chamber of the lime kiln. These particles contain, according to the analysis of thin-sections, the siliceous hydraulic components mentioned above.[16,17] However, sintered siliceous matter encapsulates and isolates lumps of burnt lime (CaO), thus preventing water access and delaying their regular slaking. This problem occurred equally with both wet-slaked and the hot lime plaster mixes.

By controlling the temperature to give softer burning, the 'pitting and popping' was reduced.[18] However, this softer burning tended to limit the variety of hydraulic components in the mortar.

Slaking process: 'wet' (to putty)

When slaking lime by the wet method, the guidelines prescribed by the old, local lime burners were followed. According to tradition, quicklime is slaked in two stages: water is sprayed over a heap of burnt limestone and, as heat is generated, the lime disintegrates into a coarse powder. As soon as the quicklime is decomposed, more water is added and the fluid putty is sieved into the lime pit.[19] Lime putty is stored in timber-lined pits, and left to mature for as long as practical, preferably for a year or more.

While digging out lime for making mortar, two forms of slaked lime can be distinguished: a conical core of heterogeneous, dark particles in the central part of the pit and at the bottom, and a lighter, creamy putty floating around it. Lime inclusions, unburnt and other particles found in the traditional, wet-slaked mortar originate mostly from this heterogeneous core. Samples of lime from this core, examined by X-ray diffraction, contain more calcium carbonate (evaluated as unburnt fragments of limestone) than samples from the creamy part.

Slaking process: 'hot'

The second method applied was the hot method – which slaked lime with sand, following instructions from nineteenth-century manuals.[20] Fresh lumps of quicklime were wetted down within a heap of sand and blended after one day. A more rational method was developed to meet the needs for larger quantities of lime for large-scale mortar production. Naturally moist sand is spread at the bottom of a wooden container, covered by a layer of lumps of quicklime and followed by a layer of sand. Water is added to dry slake the lime: double the quantity necessary for the 'theoretical' slaking itself in order to compensate for vaporization. The container is then covered (though this cover should not be airtight) for about 24 hours, in order to preserve the vapour and heat. The temperature inside a heap of

Figure 8   Slaking lime with sand, the hot method. Quicklime is spread over a sand bed.
Figure 9   The lime is covered by a layer of sand (sand–quicklime–sand 'sandwich') and sprayed with water.
Figure 10   The slaking process starts under cover of sand. The volume of lime–sand increases as the quicklime decomposes and the mixture dries before being screened.

lump lime covered by sand during slaking, can reach up to 400°C.[21] The resultant heat dries the sand, enabling the mix to be screened (screen size 4.8 mm) to remove unslaked lime and oversize aggregate. The lime–sand mix is then transferred into a roller pan mixer, wetted down and mixed to make mortar. The mortar is laid aside, stored and allowed to mature for at least a month.

In the process of hot slaking, the hydraulic reaction and the interaction between lime and sand start directly, as soon as lime is mixed with water and sand. The heat and the alkaline (basic) environment affects the surfaces of the sand grains, etching them, and improving the bond between binder and aggregate. Heat, vapour, alkaline environment and slow drying affects both the properties of the fresh mortar and the resulting plaster.

Both wet and hot slaking methods develop initial heat within the quicklime pieces. However, mortars produced from both lime types are used cold.

THE PRODUCTION OF MORTAR

Hot lime slaking and the production of mortar is a combined process. Hot mortar is mixed in a roller pan mixer (first used in full scale plastering from 2007) to ensure that the remaining, unslaked particles of lime are ground by the friction between lime particles and sand. Wet mortar is made by mixing the matured lime putty and sand in a paddle- or a roller-pan mixer. Both kinds of mortar, hot and wet, are initially mixed for one hour and then transferred to containers to mature for about a month. Both types tend to become compact and hard, partly as water is absorbed and partly due to the slowly progressing hydraulic reaction. After maturing, and before being applied on the façade, the mortars are knocked up and remixed for 30 to 60 minutes.

Both kinds of mortar are prepared with a similar content of lime, which makes it possible to compare and to evaluate the properties of the two. Batches of quicklime are measured by weight, while the wet-slaked lime is measured by volume.

Wet lime mortars, used for masonry repairs, and the coarse stuff, used for the base layers of plaster, consist of a mixture of lime–sand, ratio 1:2, by volume. The lime–sand ratio for the finishing layer of fine stuff is 1:1.5. The particle size for the well-graded sand for coarse stuff ranges from 0 to 4.8 mm, and 0 to 3 mm for the fine stuff. When hot mortar was prepared an appropriate weight of quicklime that corresponds to the fixed volume of lime putty was used; 10 litres of lime putty matured for one year corresponds to 5.3 kg quicklime (a quantity that was first calculated approximately and then adjusted on a trial and error basis). The resulting plasters (both hot and wet), with lime:sand ratio 1:2, contain approximately 43% binder and 57% aggregate, if the air content is disregarded. This was calculated through microscopy and point counting (on mortar samples cured in situ for one year). To compensate for small amounts of unslaked lime screened out during the process, a minor stock of dry-slaked lime is available.

Figure 11   Station for mortar production. Hot lime and sand are screened, transported and mixed in a roller pan mixer.

APPLICATION OF MORTAR

The final plaster consists of coarse stuff applied in several layers and one fine finishing coat. The layers of coarse mortar are applied in two ways, manually or with a spray gun (which had been in regular use in full-scale plastering since 2007). The spray gun is used for two reasons: first for economy, as the total façade area to be replastered measures over 10,000 m², and second because of an assumed shortage of skilled plasterers, the device could be operated by less experienced craftsmen. The spray gun, however, requires more water and the addition of small quantity of superplasticizer (which helps to reduce the quantity of added water). The wet 'spray-mortar' has to be applied in six thin layers (one a day) to avoid shrinkage fissures, instead of the usual two to three layers when plaster is applied by conventional manual methods. The evaluation of plaster applied by both the manual and spray gun method after one year of curing in situ showed only marginal differences; the differences between 'hot' and 'wet' mortar plasters being more significant (see 'Comparing hot mortars and wet mortars'). However, it should be noted that manual application

Figure 12   Mortar applied manually with a trowel and a float.

Figure 13   Mortar applied with a spray gun.

could probably compete economically with the mechanical spray-gun application if enough experienced craftsmen were available.

The finishing coat has to be worked up more thoroughly than the scratch coats to achieve a suitable finish. The finishing coat is therefore applied manually, thrown with a trowel and compacted with a float. To avoid lime being drawn to the surface the plaster needs to be floated and compressed when it is just at the right stage of drying, depending on the climatic conditions. Experiments using the spray-gun method for finish coats were unsuccessful because large surfaces had to be sprayed at once. This gave craftsmen insufficient time to compress the surface satisfactorily before it became too dry and hard.

Limewash was applied to the plaster surface approximately one season after the application of the mortar. It consisted of a suspension of the wet-slaked binder in water, with no added pigment. The total thickness of the plaster was 2.0–3.5 cm. The plaster surface followed the uneven form of the wall behind.

Experienced craftsmen executed the external repair and rendering, and inspected and evaluated the results.[22] According to their judgement both hot and wet mortar had good workability, good plasticity and spread easily, and both had good adhesion to the underlying stone and brick. However, the wet mortar felt leaner, was less sticky and dried more quickly, but too quickly when applied on brick. Hot mortar dried at a slower pace and retained water for longer. This favourable water-retaining capacity was an important quality, especially when mortar was applied to brick walls; the finishing coat had to retain moisture and workability while the surface was being applied and worked up with the float.

The plasterers confirmed that they preferred the hot mortars. They had a better chance of working on the finish for longer and floating and compressing the finishing layer in a satisfactory way.

COMPARING HOT MORTARS AND WET MORTARS

Methods of examination

All the finished surfaces were visually surveyed and examined in situ after they had cured for about a year. Covered scaffolding was kept in place through the first winter to protect the plaster and to facilitate survey and sampling the following spring.

Investigations in situ included evaluation of carbonation, rebound values[23] (indicating compressive strength), tendency towards water suction and moisture conditions in the plaster.[24] Samples of the different plaster types cured in situ for one year were cut out and examined in the laboratory with mechanical tests of their physical properties such as density, porosity, water absorption. Properties such as structure, binder–aggregate–air void ratio, carbonation depth, binder related particles and fissures were studied by thin-section microscopy.[25,26] The majority of the laboratory analyses were conducted on approximately 5–10 cm^2 samples cut out from the castle's façade.

The investigation/test methods used were chosen for pragmatic reasons, because they could be adapted to the conditions at the building site, and consisted of simple tests in the laboratory. Testing and experiments were conducted in parallel with the ongoing works, after the new plaster has been applied and cured. The results and conclusions could only be implemented in the following season; which meant a two-year lag.

Structure in general

The investigation of the mortar structure was based, to a large extent, on the analysis of thin sections and point counting. The 'hot' as well as 'wet' mortar plasters prepared at Läckö, have a complex internal structure, quite different from the structure of modern, commercial lime mortars, which are more homogeneous in general. The structure of hot and wet plasters at Läckö is also different from that of the extremely binder-rich original, historic mortars (Plates 23 and 24).

The matrix consists of microcrystalline lime with crystals of 1–3 μm, and contains different quantities of undispersed lime-related particles and residues of unburned limestone (see 'Lime burning' and 'Lime inclusions and other binder-related particles'). Red-brownish particles of ferrous clay compounds (up to 20 μm) appear both in the matrix and in the particles with hydraulic properties. Preliminary SEM-analysis[27] showed fused, interpenetrated, fine, almost needle-shaped lime crystals.[28] No difference was observed between the shape of crystals in plasters made of hot or wet lime.

Images of thin-sections of both hot and wet plaster samples (mortars with lime–sand ratio 1:2, by volume) show sand grains packed tightly together and glued by lime-binder; the aggregate acting as a 'skeleton'. When the lime–sand ratio is increased to 1:1 by volume, a more historic compact structure appears, with sand grains distributed more evenly in the matrix. They are isolated from one another, almost as in the fat, historic mortars. Air voids observed in the lime-rich matrix have a more distinct, round form and are disconnected from sand grains. In general, there are fewer but coarser micro-fissures in these lime-rich mortars, both in the historic original and the new versions. The experimentally reproduced plaster samples with lime–sand ratio 2 : 1 or 4 : 1 have a typical historic structure.

Carbonation and frost resistance

To date, no damage has been observed on the façades; either on the 'conventional' plaster produced from lime putty or on the hot mortar plaster. Fine shrinkage fissures were observed before the application of the limewash, but the lime paint binder filled these. One of the most important observations was that there were no signs of frost damage. Six years, however, is too short a period of time for long-term evaluation of the external plaster surfaces on a historic building, and this should therefore be regarded as an interim evaluation.

In general, plaster made of hot mortar is, after one year of curing in situ, slightly harder than plaster made of corresponding putty mortar, despite the slower carbonation of the hot mortar. Comparative measurements of rebound values, which can be correlated to compressive strength, have been undertaken with a Concrete Test Hammer.[29]

Carbonation progresses at a slower rate in hot mortar plasters than in the wet ones. Phenolphthalein tests were carried out in situ after one year of curing. The wet plaster was in general more deeply carbonated than the hot plaster. This carbonation rate could also be observed on the thin-section samples cut out from the façade after one year of curing. All the samples of wet mortar plaster applied on the northern façade in 2007 (collected in 2008) were completely carbonated while of the six samples of hot mortar plaster taken, only one was. The slower carbonation constituted a disadvantage that was compensated for in the Läckö project by protecting the plaster during the first winter. The plaster thus had a chance to carbonate in depth before it was exposed to freezing conditions.

The laboratory examination of frost resistance proved to be impracticable on mortar samples cut out from the façade because of varying thicknesses and other factors. However, the results obtained on the façades in severe climatic conditions and repeated temperature transitions (around 0°C) during the winters of 2003–09 seem to be more reliable, and no frost damage has been observed on any of the newly plastered areas.

Rebound value

Figure 14   Comparison of rebound values on wet and hot mortar plaster. The dot and square indicate average rebound values for mortars applied in 2004, tested/evaluated in 2005. Black lines indicate the span, max–min of all rebound values measured on each surface. High value means hardness and is interpreted as high compressive strength.

Density, porosity and air content

Bulk density and porosity were determined from samples cut out from the plaster surfaces after one year of curing. Porosity was determined in two stages: first after immersion in water for 24 hours and then after the subjection of the same samples to forced vacuum under water for one hour. The porosity measured after 24 hours is called partial porosity and the porosity in samples subjected to both stages is called total porosity. The calculated difference between the total and partial porosity constitutes closed porosity, a value that corresponds roughly to the volume of encapsulated air voids observed in thin-sections.[30] It was assumed that the encapsulated air voids would only take up water slowly and would constitute a buffer zone into which water could expand when freezing.

Tests were undertaken in situ to observe the absorption of water. Water was sprayed onto comparable plaster surfaces and it was seen that wet mortar plaster absorbed water more readily than hot mortar plasters. Consequently, it could be expected that hot plaster would absorb rainwater more slowly than wet mortar plasters. The plasterers at Läckö also found

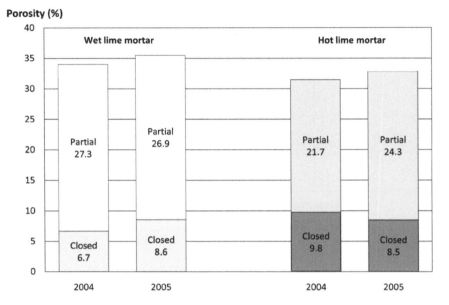

Figure 15   Comparison between total porosity, partial porosity and closed porosity in hot and wet mortar plasters with lime–sand ratio 1:2. Average values from plaster surfaces applied in 2004 and 2005.

that the hot plaster retained water for longer. These observations were confirmed by the physical tests in laboratory (bulk density and porosity) and the subsequent microscopy of thin-sections carried out on neighbouring samples[31] from plaster surfaces applied in 2004 and 2005 and tested in the following years 2005 and 2006.

Tests of bulk density and porosity carried out in the years 2004 and 2005 (and tested in the following years 2005–06) indicated that:

- hot mortar plasters had on average slightly higher density than wet mortar plasters;
- hot mortar plasters had on average lower total porosity and partial porosity, but plasters made of hot mortar had on average a higher value of closed porosity than the ones composed of wet mortar; i.e. the volume of encapsulated air voids was higher in hot mortars.

However, the results from the examination of density and porosity undertaken on wet and hot plaster samples from the northern façade in 2008 (replastered in 2007) were ambiguous.[32] It was thus difficult to give priority to either plaster type, hot or wet. A plausible reason for the different test results, compared to the earlier ones, could be found in the slow, delayed carbonation of the hot mortar on the northern façade.

The assumption that a larger amount of enclosed air voids promotes frost resistance and that hot lime mortar (eventually) possesses this quality still has to be confirmed by new tests.

In general terms, the higher the lime content in the mortar, the higher the porosity, both total and partial. However, the higher the lime content in the mortar, the lower the difference between the total and partial porosity, i.e.: closed porosity. Decreasing closed porosity arises in plasters with increasing lime content. This was confirmed by the tests carried out on series of the newly produced historic plaster samples,[33] both hot and wet. Microscopy of thin sections (followed by point counting) also confirmed these observations (Plates 25–28).

When the extremely lime-rich samples of the newly produced 'historic plaster' with different lime content were examined by microscopy of thin-sections, only a few air voids were found. These have a distinctly rounded form, very much like the air voids in the original, historic mortars at Läckö, and quite unlike the larger quantities of air voids with irregular form that prevail in the ordinary lime mortars with lime–sand ratio 1:2 (Plate 29).

Lime inclusions and other binder-related particles

The binder in plaster samples cured in situ for one year was examined by microscopy. Binder-related particles were quantified by point counting. The binder consisted of a primarily homogeneous microcrystalline matrix with a variety of solid particles:

- inclusions of undispersed lime constitute up to 22% of the total volume of the binder;
- particles of incompletely burned lime constitute up to 10% of volume of the binder;
- particles with hydraulic properties constitute up to 6% of volume of the binder, and
- traces of alum shale constitute less than 1% of the volume of the binder.

The quantities above refer to mortars applied in 2004 and 2005 and cut out from the façade after one year of curing.

These binder-related particles originate from the limestone nature and source, and also to some extent from the methods applied in lime burning and slaking. Mixing methods when producing mortar also affect, to some degree, the quantity of lime inclusions in all types of mortars. Analysis of the different particles may contribute to our understanding of the properties of mortars, and may also be utilized as a basis for comparison with historic mortars.

Lime inclusions

Inclusions of undispersed lime have predominantly non-hydraulic properties. They have, in general, a lighter colour than the surrounding matrix, are relatively porous, and measure up to 2 mm. They are considered to act as an aggregate. Lime inclusions constitute:

- in hot mortars: 0–6% of the total volume of the binder (2% on average);
- in wet mortars: 5–22% of the total volume of the binder (12–13% on average).

The values are consistent for all mortar samples from the year 2004–05 tests, which were mixed and remixed for at least 1.5 hours in total.

Lime inclusions are found primarily in the samples of wet mortars and are less frequent in the samples of hot mortars. The difference between the quantities of lime inclusions in wet and in hot mortar is substantial. Efficient mechanical mixing for a long period of time (more than 1.5 hours) in a paddle or a roller-pan mixer makes only a minor impact on the quantity of lime inclusions in the wet mortars, and a substantial impact in hot mortars.

The slaking process presumably affects the quality, as well as the quantity, of lime inclusions. Inclusions in hot mortars could be weakened by hot slaking (by the generation of heat, hot vapour and drying) and thus be easily destroyed by mechanical mixing. Poorly mixed hot mortar could retain a larger quantity of these weakened lime inclusions, while well-mixed hot mortar contains hardly any. Original historic mortars have larger quantities of lime inclusions than the mortars produced at Läckö today, probably because they were not mechanically mixed.

Particles of incompletely burnt lime

Particles of incompletely burned lime constitute up to 10% of the total volume of binder (on average 3%). In general these particles are less frequent in the original, historic mortars, where they constitute 4% at most of the volume of the binder. This confirms the assumption that the burning temperatures today are lower, and/or possibly the process of lime burning is shorter, than when the original historic lime was burned (see 'Lime burning') (Plate 30, 31).

Particles with hydraulic components

Particles that contain hydraulic components derived from clay minerals and siliceous matter in the original limestone, when hydrated, affect the hydraulic properties of the binder. The particles are clearly distinguishable in the thin-sections. The largest particles in the thin-sections measure 1.9 mm in diameter. A dark, condensed zone of binder that has lower pore size than the surrounding binder encircles many of the particles. Wherever the hydraulic particles have contact with air pores, needle-like crystals are formed on their surface (probably hydrate of calcium-aluminates or silicates).

The frequency of particles with hydraulic properties depends on the presence of the hydraulic components in the limestone, and on the temperature achieved during burning of the limestone. These factors were similar in the two types of mortar, the hot and the wet, and were not influenced by the slaking process.

The quantity of hydraulic particles in the samples composed of hot and wet mortar appeared to be similar. The quantity varied in the hot mortars from 0 to 7% of the volume of the binder, while in the wet mortars it varied from 2 to 8%. The resulting hydraulicity in both types of mortar was estimated as low. The hydraulic properties in this type of feebly hydraulic lime only developed slowly.

Traces of alum shale that are found in the mortar originate from limestone that was quarried from limestone beds interlayered with alum shales (see 'Materials'). The particles of alum shale are burnt together with the limestone and show a characteristic red colour after being burnt. They are an integral part of the matrix and participated in the hydraulic reaction.

CONCLUSIONS

Two types of mortar are compared in this study: mortar made of hot lime and of wet lime. The raw materials, limestone and sand, came from common, local sources. The limestone for each test series was burnt on the same occasion. Mortar production was differentiated first when slaking lime. The different slaking procedures resulted in variations to the mixing, sieving and storing methods required. The application techniques were common: spray gun for the inner coats of mortar and manual application for the finishing coat. The properties of the hot and wet mortars vary. The study of long-term properties of hot mortar as compared to wet mortar has to be continued.

Looking at the plastered façade following completion, only an experienced plasterer would be able to note the difference between hot and wet mortar, and then only when the mortar is still green. The noted differences apply primarily to the properties of the fresh mortars and the following application methods: 'hot mortar feels fatter' and 'hot mortar can be worked and compacted for longer'.

Both mortar types, hot and wet, possessed acceptable properties. However, the practical advantages of hot mortar provided the main incentive for the proprietor–manager, National Property Board, to choose the hot method:

- shorter production process: from lime burning to the ready plaster on the wall;
- no need to burn lime one year in advance and to construct lime pits;
- good workability: easy to apply and work up manually;
- good physical properties.

The disadvantages of hot lime compared with wet lime:

- more complicated logistics of production; joint process of slaking lime and mixing mortar;
- facilities for storage of the prefabricated mortar are required;
- slower carbonation than plaster made of wet lime and consequently longer protection of the fresh plaster required while curing.

The objective of producing compatible mortar has been partly fulfilled by using binder of local limestone and applying traditional methods of burning and slaking. The ambition to replicate the formulations of the original, historic, binder-rich mortars was abandoned for technical/practical reasons and for economic reasons (expense of lime manufacture). It remains uncertain whether hot, wet or other slaking methods had been used in the past and further research is needed to provide a comparison between the properties and application methods of newly produced mortars and the original historic mortars.

For plates referred to in this paper, see the colour section following page 132.

This paper was first published in the *Journal of Architectural Conservation*, Volume 17, No. 1, March 2011.

Notes

1 Jonsson, L. (ed.), *Läckö; Landskapet, borgen, slottet (Läckö; the Landscape, the Fort, the Castle)*, Carlssons Bokförlag, Stockholm, 1999.
2 Examination methods: building archaeology and microscopy of thin-sections.
3 Sandström Malinowski, E., 'Rediscovering historic mortars and skills at Läckö Castle', *The Journal of the Building Limes Forum*, Vol. 12, 2005, pp. 9–21. Sandström Malinowski, E., 'Historic mortars revived: Developing local materials and crafts for restoration', *Repairs Mortars for Historic Masonry*, RILEM Publications SARL, 2009, pp. 328–38.
4 Gibbons, P., *Preparation and Use of Lime Mortars*, Scottish Lime Centre. Historic Scotland, Edinburgh, 2003.
5 Forster, A., 'Hot lime mortars: A current perspective', *Journal of Architectural Conservation*, Vol. 10, No. 3, 2004, pp. 7–27.
6 Arcolao, C., *Le ricette del Restauro. Malte, intonaci, stucchi dal XV al XIX secolo*. Saggi Marsilio, Venezia, 1998.
7 The lime slaked in a pit under a thick layer of sand benefits from the heat and the vapour developed; the method is referred to as 'slaking by sprinkling'.
8 Rothstein, E. E. von, *Handledning i allmäna byggnadsläran (Manual for Building Instruction)*, 3rd edition, Stockholm, 1856, 1873, 1890.
9 Holmes, S., 'Hot lime in a cold climate', *Lime News, The Journal of the Building Limes Forum*, Vol. 1, No. 2, 1993, pp. 19–29.
10 Lindqvist, J. E., Petrografisk analys av kalksten från Kakeled (Petrografic Analysis of Limestone from Kakeled) unpublished report, SP Technical Research Institute of Sweden, 2005.
11 Davey, N., *A History of Building Materials*, Phoenix House, London, 1961.
12 Kiln temperatures are variable in traditional and experimental kilns.
13 Sandström Malinowski, E., Historic Mortars at Läckö Castle, in prep, 2011. Sandström Malinowski, E., Interim reports: PS-3, PS-4, PS-5, PS-7, unpublished, 2004–09.
14 Seir Hansen, T., Research project reports, Microscopy of thin sections – historic/original mortars, 2002–09.
15 Seir Hansen, T., Research project reports, Microscopy of thin sections – contemporary mortars, 2005–09.
16 Sandström Malinowski, Historic Mortars, in prep 2011.
17 Seir Hansen, *op. cit.*, 2005–09.
18 It is finally reduced by sieving the lime after slaking and by efficient mixing of lime and sand in a roller pan mixer.
19 The sieve used has round apertures with a diameter of 5 mm.
20 Rothstein, *op. cit.*
21 Measured by a temperature sensor, Technoterms 9300.
22 Zälle, J. and Zälle, P., Works Managers Läckö Castle, Informants 2003–2008.
23 Concrete Test Hammer Type N (PROCEQ N-34).
24 Exotec Instrument, MC-160SA.
25 Sandström Malinowski, Historic Mortars, in prep 2011.
26 Seir Hansen, *op. cit.*, 2005–09.

27 JEOL, 5310LV Scanning Electron Microscope. Samples of uncoated mortar fragments.
28 Seir Hansen, *op. cit.*, 2005–09.
29 Concrete Test Hammer Type N (PROCEQ N-34).
30 Each sample has been split into two parts: one part is used for the examination of porosity by immersion, the second by microscopy of thin sections followed by point counting.
31 Seir Hansen, *op. cit.*, 2002–09.
32 Nearly equal density in hot and wet plasters and slightly higher volume of 'enclosed' air voids in the wet plasters.
33 The newly produced samples of 'historic plaster' have lime–sand ratios 1:2, 1:1, 2:1 and 4:1.

A Study of Historic Lime Mortars at Ardfert Cathedral

Grellan D. Rourke

INTRODUCTION

This research forms part of an ongoing project, begun in the 1990s, to gather detailed information on historic mortars and to use this information, in the first instance, to specify appropriate mortars for the conservation and repair of historic buildings. As the Heritage Service undertakes large-scale works to such buildings the historic mortars are analysed in order to devise suitable recipes. These recipes will eventually form a database from which additional research will evolve to examine the development of lime mortars in Ireland over time and to demonstrate how their use may have spread. It will also help to inform us about technologies which existed in earlier ages. This overall project is long-term: only when a sufficiently large database is assembled can its full potential be realized.

This project has been multidisciplinary in its approach. Initially, investigation took place on site at a practical level with the purpose of choosing suitable mortar and plaster samples. The analytical results were developed with the use of science and technology to elicit the type of information required to devise recipes for the recreation of replica historic mortars to be used in the repair of the cathedral. This will have application for use on other similar projects. Detailed specifications were then derived and materials were sourced and modified to suit the requirements. Practical experimentation was then undertaken on site to investigate the process of mixing to achieve quality mortars and plasters with excellent workability. Training was undertaken to maximize the standard of workmanship and supervision was also an integral element in achieving good, consistent results.

DESCRIPTION OF THE CATHEDRAL

Ardfert Cathedral is located in the village of Ardfert in County Kerry. It lies within a few miles of the coast and the prevailing winds are south-westerly. Great sand dunes line the coast at this point and the strong winds carry the sand particles to the site. This action has caused considerable erosion, particularly to the vulnerable dressed and carved sandstone on the south and west elevations. Ardfert Cathedral dates from a multiplicity of periods, earlier structures being incorporated into a thirteenth-century cathedral which was then altered and added to in subsequent centuries. Walls date from almost every century from the eleventh through to the nineteenth, and the cathedral is probably unique in Ireland in hosting such a broad range of construction periods.

The main body of the cathedral, built in the thirteenth century, is like an elongated shoebox in shape, with a south aisle at the west end. This cathedral incorporates the north wall of the nave of an eleventh-century church, and the remains of the twelfth-century west door of an earlier Romanesque church. A sacristy was then added and, in the fifteenth century, a south transept was constructed. At this time the cathedral was much enhanced. A new roof structure was constructed, allowing the entire building to be battlemented, raising the height of the walls half as much again and creating a very imposing structure. Later again, a two-storey priest's house was constructed at the north-east corner. In 1671 the south transept was extended to create a Church of Ireland parish church. An east chapel was added to this church in the nineteenth century.

With the exception of the south transept the cathedral had been roof-less since the late sixteenth century, and since that time there had been a continual slow deterioration. The action of the weather had consider-ably reduced the strength of the walls, in places washing out the mortar and creating voids within. Elsewhere, the mortars had lost considerable tensile strength due to loss of binder and were no longer holding the walls together in places where they were under considerable pressure. Large burials constructed up against the walls internally and externally had undermined foundations and the south wall was in quite a precarious structural condition. During the nineteenth century there had been two interventions to prevent this wall from collapsing outwards. When works

Figure 1   General view of Ardfert Cathedral before commencement of works (courtesy of Con Brogan, DAHG).

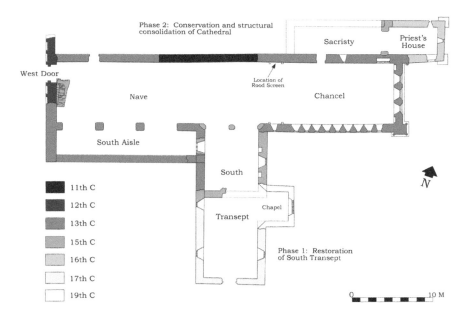

Figure 2   Plan of Ardfert Cathedral showing the dating sequence.

were undertaken here in the late nineteenth and early twentieth centuries inappropriate mortars were used, and these caused serious damage to the softer decorative sandstones.

The main body of the cathedral is constructed of local limestone, which was quarried nearby on the north side of the site. Local sandstones were used for the Romanesque west door, and most of the door and window elements and decorative details of the thirteenth-century construction. The sandstones used are of varying hardness and porosity and present in a wide range of colours and shades. The style of masonry varies considerably throughout; one can trace the development of local masonry construction through the centuries on this site.

THE OVERALL PROJECT

This major project was initiated in the local community and later, once suitable funding became available, it became part of a major programme of works to National Monuments by the Office of Public Works. Initially, between 1989 and 1995, an extensive archaeological excavation took place and burials within and close to the cathedral were removed and re-interred in the graveyard alongside. The work to the fabric of the building was undertaken in two separate stand-alone phases, both of which were carried out by contract. Phase 1, the restoration of the south transept, was completed in 1994, and this area serves as a reception area for visitors. A guide service is based here, and there is an exhibition of the history of the site and a display of the most interesting decorative stones recovered in the excavation. At the same time another small project was undertaken. The Romanesque west doorway posed very particular conservation problems and it was decided that the work here was unsuited to being contracted out. This work was carried out by the Office of Public Work's direct labour workforce, with specialist help.

There was a considerable time lag before Phase 2 of the project commenced, and one aspect of the work gave cause for concern. The excavation had exposed the entire base of the cathedral walls, and it became clear that the exposed foundations would need more immediate attention. The walls required underpinning and it became necessary to complete this work prior to the commencement of Phase 2. This work was undertaken by a specialist contractor.

Phase 2 comprised the structural consolidation of the cathedral itself, including the stabilizing of the very unstable east end of the south wall, the grouting of the walls with a lime-based grout, the repair of the south-east corner and the battlements, and full conservation of the remainder of the cathedral fabric. This work involved a main contractor and a series of specialist subcontractors. Phase 2 started on site in July 1999. The duration of the project was approximately sixteen months, and it involved a

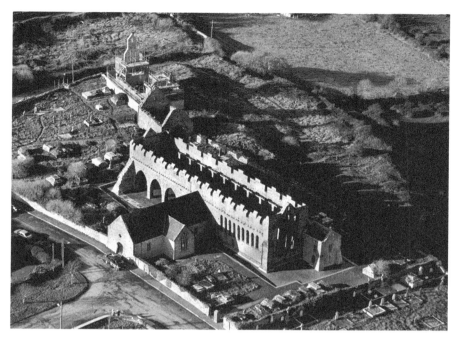

Figure 3 Aerial view of Ardfert Cathedral from the south-east after works. The quarry lies in the hollow to the north of the cathedral (courtesy of Con Brogan, DAHG).

core workforce of up to twelve men with a full-time clerk of works. This figure increased when scaffolders and the various subcontractors came onto the site. The contract cost was €1.9 million.[2] The scale of Phase 2 and the lead-in time gave us the opportunity to study the historic mortars in great detail and to recreate new mortars to match the existing ones for use in the repairs.

THE HISTORIC MORTARS

Sampling of historic mortars

Before any sampling was undertaken a full range of mortars was identified from the different periods, and also from within individual periods where there were both general wall mortars and fine mortars for bedding tight-jointed ornamented sandstone. There are remains of internal plasters dating from the thirteenth and fifteenth centuries and these were

Figure 4   The interior of the east wall of the cathedral where AC07 was taken. Two samples were removed, marked (a) and (b). These samples were taken in March 1998 (courtesy of Grainne Shaffrey).

included in the study. Samples (AC01–AC21) were extracted from within the walls to avoid any contamination from later repointing, and from areas which did not show any later masonry repairs. More than one sample was taken for each period so that a better statistical range of mortar could be included. These samples were then submitted to conservation specialists Carrig for analysis.

Analysis of the mortars and plasters: methodology[3]

The aim was to assess the nature, composition, microstructure and condition of the mortars and plasters from Ardfert Cathedral. The samples were analysed by both physical and chemical methods. All samples contained loose material as well as compact fragments.

The samples were carefully sieved and the compact fragments and loose matter were studied separately. The loose aggregate was washed in distilled water to remove the remaining lime binder and organic matter. Once dry and clean, the fragments of mortar and plaster were studied using a magnifying glass and a stereomicroscope. A selection of these fragments was exposed to attack by standard hydrochloric acid digestion to assess the nature and composition of some aggregate and binder. When the visual examination and sieving had taken place, the results were set out in a table giving a brief description of the sample, its components and its condition.

Based on this initial study, specimens were chosen for thin-section preparation. The thin sections were analysed by optical methods with a petrographic microscope, using both parallel and polarized transmitted light. The selection of the fragments was based on two criteria: the variability of their components, and their location in the cathedral. A detailed petrographic description of each sample was made.

The results of the stereomicroscopic examination were also set out in a table under the headings: aggregate, other components, matching of samples, and decay. An example is given below (AC15: a thirteenth-century sample from the north wall).

Sample	Aggregate	Other components	Matching mortars	Decay
AC15	Course aggregate of sub-rounded sandstone, shale, quartzite and basic igneous rocks sized 9–28 mm (approx. 3% in volume) Angular limestone aggregate sized between 1 and 5 mm	Lime binder Organic debris Lime lumps average sized 2–3 mm	ACO6 ACO8	Weathered but still relatively compact Surface dissolution and granular disintegration

Results of analysis

The mortars analysed were all lime-based mortar mixes. They were composed of fine, medium and coarse aggregate cemented with a lime binder.[4] The coarse aggregate consisted of angular fragments of limestone as well as sub-rounded and rounded fragments of chert and different types of sandstone. The medium-sized aggregate included sub-rounded fragments of laminated shale, quartz grains and fragments of sandstones. The fine aggregate was usually quartz, fragments of shells, calcite grains and fossils.

The plasters were found to be lime-rich mixes containing aggregate of variable size, nature and composition. Two different plasters were identified, which corresponded with the thirteenth- and fifteenth-century plasters submitted for study.

The chemistry and mineralogy of the samples was derived from petrographic analysis. The main chemical elements contained in the samples were $CaCO_3$, SiO_2, Al_2O_3, Na_2O, Fe_2O_3, K_2O and trace elements. The main chemical element is calcium, as both plasters and mortars largely consisted of limestone aggregate ($CaCO_3$) and carbonated lime ($Ca(OH)_2$) binder. Silicon is the main component of sandstone and quartzite aggregate. The minerals contained in the samples fell into three main groups: minerals in the aggregate, minerals in the binder, and minerals in the loose fine detrital fraction.

The analytical investigation showed that all the historic mortars, with one exception, were mildly hydraulic. Small quantities of clay and ash were found in the samples and these would have given the mortars a low level of hydraulicity. It is interesting that the clay content is static, about 1% (see 'The burnt-clay component', below). There were also 'additives' such as shell, charcoal, and chert.

Devising mortar recipes to match the historic mortars

The analytical work undertaken provided scientific information on the make-up of the mortars. However, this was not immediately user-friendly or accessible from a building construction perspective. The chemical make-up information, as presented, did not translate easily into mortar components: more tangible information was required, such as the full size range and type of aggregates. Discussion with Dr Pavia led to some further work being undertaken to set out the information in such a way that it could be used for the purpose of recreating the full range of historic mortars.

The next step was to set about devising suitable recipes for the replica historic mortars.[5] A chart was devised for each mortar, setting out the basic mix: the variety and size range of aggregates to be used as a percentage by volume, and details of the additives. At the base of each chart a recipe was set out employing units of measurement which could be easily used by those on site to mix the mortars in advance.

The five samples illustrated below give a comparison between the eleventh-, thirteenth- and fifteenth-century mortars and the thirteenth- and fifteenth-century plasters.

Mix B – 11th Century Mortar(AC01)

LIME   : SAND/
PUTTY   AGGREGATE
1   : 3

Constituents

			Remarks
A1 Quartz Sand	90% vol. of sand/aggregate	Size range: 40µm ≥ x ≤ 2.0mm	No greater than 6% should be below 60µm. Architects may substitute alternative sieve sizes.
		Sieve sizes BS 410	
		Ref. Table A	
A2 Limestone Sand	2% vol. of sand/aggregate	Size range: 4mm ≥ x ≤ 18mm	Coarse. Predominately in the lower end of range. Architects may substitute alternative sieve sizes.
		Sieve sizes BS 410	
		Ref. Table A	

Sand shall be sharp, well graded, angular and free of soil, organic matter, soluble salts and other impurities.

Additives

			Remarks
Shells	5% vol. of sand/aggregate		Washed, unevenly crushed to Architect's approval.
Burnt Clay	1% vol. of sand/aggregate	Polestar501	
Charcoal	2% vol. of sand/aggregate	Size range: x ≤ 1mm	Charcoal shall be crushed to Architect's approval.
		Sieve sizes: BS 410	
		1mm	

Additives shall be added to sand/aggregate mix

SUMMARY

LIME	:	SAND/AGGREGATE				
1	:	3				
Lime		Quartz Sand	Limestone Sand	Shells	Burnt Clay	Charcoal
100 units		270 units	6 units	15 units	3 units	6 units

Mix C – 13th Century Mortar(AC06)

LIME   :   SAND/
PUTTY :   AGGREGATE
1   :   3

Constituents

			Remarks
A1 Quartz Sand	80% vol. of sand/aggregate (75% if shells added)	Size range: 1mm ≥ x ≤ 5mm Sieve sizes BS 410 Ref Table A	Well graded, sharp & angular. No greater than 6% should be below 60μm. Sizes may change.
A4 Sandstone Sand	14% vol. of sand/aggregate		
A5 Sandstone Gravel	3% vol. of sand/aggregate	Size range: 4mm ≥ x ≤ 18mm Sieve sizes BS 410 Ref Table A	Coarse, sub-rounded. Predominately in the lower end of range. Architects may substitute alternative sieve sizes.

Sand shall be sharp, well-graded, angular and free of soil, organic matter, soluble salts and other impurities.

Additives

			Remarks
Shells	5% vol. of sand/aggregate		Washed, unevenly crushed to Architect's approval.
Burnt Clay	1% vol. of sand/aggregate	Polestar501	
Charcoal	2% vol. of sand/aggregate	Size range: x ≤ 1mm Sieve sizes: BS 410 1mm	Charcoal shall be crushed to Architect's approval.

Additives shall be added to sand/aggregate mix.   *Shells to be added to Architects instruction only.

SUMMARY

LIME		SAND/AGGREGATE					
1 Lime		3 Quartz Sand	Sandstone Sand	Sandstone Gravel	Burnt Clay	Charcoal	Shells
100 units	:	240 units	42 units	9 units	3 units	6 units	
or 100 units	:	225 units	42 units	9 units	3 units	6 units	15 units

Mix D – 15th Century Plaster (AC09)

LIME	:	SAND/	
PUTTY	:	AGGREGATE	
1	:	3	

Constituents

			Remarks
A1 Quartz Sand	43% vol. of sand/aggregate	Size range: 40μm ≥ x ≤ 1.5mm	Architects may substitute alternative sieve sizes.
		Sieve sizes BS 410	
		Ref. Table A	
A2 Limestone Sand	43% vol. of sand/aggregate	Size range: 2mm ≥ x ≤5mm	Well graded, sharp & angular. No greater than 6% should be below 60μm.
		Sieve sizes BS410	
		Ref. Table A	Architects may substitute alternative sieve sizes.
A3 Limestone Gravel	1% vol. of sand/aggregate		
A4 Sandstone Sand	10% vol. of sand/aggregate	Size Range: 1mm ≥ x ≤ 5mm	Rounded to sub-rounded. Architects may substitute alternative sieve sizes.
		Sieve sizes: BS 410	
		Ref. Table A	
A5 Sandstone Gravel	2% vol. of sand/aggregate		

Sand shall be sharp, well graded, angular and free of soil, organic matter, soluble salts and other impurities.

Additives

		Remarks
Burnt Clay	1% vol. of sand/aggregate	Polestar501

Additives shall be added to sand/aggregate mix

SUMMARY

LIME	:	SAND/AGGREGATE						
1	:	3						
Lime		Quartz Sand	Limestone Sand	Limestone Gravel	Sandstone Sand	Sandstone Gravel	Burnt Clay
100 units	:	129 units	129 units	3 units	30 units	6 units	3 units

Mix I₁ – 13th Century Plaster (AC12)

LIME	:	SAND/
PUTTY	:	AGGREGATE
1	:	2½

Constituents

			Remarks
A1 Quartz Sand	48% vol. of sand/aggregate	Size range: 40µm ≥ x ≤ 2.0mm	Sharp, well-graded & angular. Architects may substitute
		Sieve sizes BS 410	alternative sieve sizes.
		Ref. Table A	
A2 Limestone Sand	45% vol. of sand/aggregate	Size range: 1mm ≥ x ≤ 3mm	Sharp well-graded & angular. No greater than 6% should be
		Sieve sizes BS 410	below 60µm.
		Ref. Table A	

Sand shall be sharp, well graded, angular and free of soil, organic matter, soluble salts and other impurities.

Additives

			Remarks
Shells	5% vol. of sand/aggregate		Washed, unevenly crushed to Architect's approval.
Burnt Clay	1% vol. of sand/aggregate	Polestar501	
Charcoal	1% vol. of sand/aggregate	Size range: x ≤ 1mm	Charcoal shall be crushed to Architect's approval.
		Sieve sizes: BS 410	
		1mm	

Additives shall be added to sand/aggregate mix

SUMMARY

LIME	:	SAND/AGGREGATE				
1	:	2½				
Lime		Quartz Sand	Limestone Sand	Shells	Burnt Clay	Charcoal
100 units		120 units	112½ units	12½ units	2½ units	2½ units

Mix I₂ – 15th Century Plaster (AC14)

LIME PUTTY	:	SAND/AGGREGATE
1	:	1½

Constituents

			Remarks
A1 Quartz Sand	20% vol. of sand/aggregate	Size range: 40μm ≥ x ≤ 2.0mm	Angular river sand. Predominantly in the lower end of range. 1mm ≥x≤ 4mm. Architects may substitute alternative sieve sizes
		Sieve sizes BS 410	
		Ref. Table A	
A2 Limestone Sand	5% vol. of sand/aggregate		
A4 Sandstone Sand	69% vol. of sand/aggregate		
A5 Sandstone Gravel	3% vol. of sand/aggregate	Size Range: 40mm ≥ x ≤ 11mm	Rounded siliceous. Predominantly in the lower end of range. No greater than 6% should be below 60μm. Architects may substitute alternative sieve sizes.
		Sieve sizes: BS 410	
		Ref. Table A	

Sand shall be sharp, well graded, angular and free of soil, organic matter, soluble salts and other impurities.

Additives

			Remarks
Shells	2% vol. of sand/aggregate		Washed, unevenly crushed to Architect's approval.
Charcoal	1% vol. of Sand/aggregate	Size range: x ≤ 1mm	Charcoal shall be crushed to Architect's approval.
		Sieve sizes: S 410	
		1mm	

Additives shall be added to sand/aggregate mix

SUMMARY

LIME	:	SAND/AGGREGATE				
1	:	1½				
Lime		Quartz Sand	Limestone Sand	Sandstone Sand	Sandstone Gravel	Shells	Charcoal
100 units		30 units	7½ units	103½ units	4½ units	3 units	1½ units

Sourcing materials

As this work was progressing the basic constituents of the mortars were examined to determine where the sands and aggregates might have originated.[6] Sourcing these materials was undertaken by Carrig and samples from various quarries in the area were gathered and examined. A sea sand from nearby Tralee Bay was an excellent match for the thirteenth-century mortars. However, it was decided not to use this due to salt contamination, and indeed obtaining permission to remove sand from a beach would not have been an easy task. Sara Pavia confirmed that there was very little difference between this and the sand from Rangue Pit near Killorglin, which was chosen for the project.

Below is Table A, which sets out the sources for the full range of aggregates to be used in the making of the new mortars.

Table A   Material Sources.

Aggregate		Source
A1	Quartz Sand	Killarney Pit, nr Killarney
A2	Limestone Sand	Ardfert Quarries, Ardfert
A3	Limestone Gravel	Ardfert Quarries, Ardfert
A4	Sandstone Sand	Rangue Pit, nr Killorglin
A5	Sandstone Gravel	Rangue Pit, nr Killorglin
A6	Chert Gravel	Doneraile, Co. Cork

Inevitably, the materials available from these sources were not fully suitable and required modification. A sieve analysis and void ratio test were carried out on the aggregates and the data obtained was tabulated and compared with the replica mortar requirements.[7] A specification was then devised using a series of sieves (to BS 410) to meet the requirements. Table B shows the full range of aggregates and the percentage blending of the secondary aggregates (SA1, SA2 etc.) to form the primary aggregates (PA1, PA2 etc).

Table B   Aggregates

	Aggregate Type	A1 Quartz Sand	A2 Limestone Sand	A3 Limestone Gravel	A4 Sandstone Sand	A5 Sandstone Gravel	Ref. Sample analysis
Mix A1 11th C Stage 1 pointing	S1	90%	None	2%	None	None	AC01
Mix A2 13th C Stage 1 pointing	S2	80%	None	None	14%	3%	AC06
Mix A3 15th C Stage 1 pointing	S3	43%	43%	1%	10%	2%	AC09
Mix B 11th C mortar	S1	90%	None	2%	None	None	AC01
Mix C 13th C mortar	S2	80% (75% if shells used)	None	None	14%	3%	AC06
Mix D 15th C mortar	S3	43%	43%	1%	10%	2%	AC09
Mix E 15th C Fine pointing	S8	43%	43%	1%	10%	2%	AC20
Mix F 13th C Fine pointing	S4	47%	25%	None	25%	None	AC04
Mix G Tooled holes	S5	50%	50%	None	None	None	AC12
Mix H Putlog	S5	50%	50%	None	None	None	AC12
Mix I$_1$ 13th C plaster	S6	48%	45%	None	None	None	AC12
Mix I$_2$ 15th C plaster	S7	20%	5%	None	69%	3%	AC14

Aggregate mixes are to be mixed thoroughly before use. Additives shall be added after aggregates have been mixed except **clay** to be added before use and this to be evenly mixed.

2no. Samples of each sand will be required for visual inspection by the Architect.

Note: Table B + Table C = Filler (100%)

'Additives' were crushed shells, charcoal and burnt clay. It was necessary to obtain permission from Kerry County Council to gather shells from the foreshore, and this was granted given the special nature of the project. These were thoroughly washed in clean water, then crushed to the appropriate size and then washed again. A charcoal source was established and the charcoal was broken into small particles to match that of the original mortars. The burnt-clay component was achieved by using Polestar 501. See Table C for constituent percentages of additives for blending with the primary aggregates (PA1, PA2 etc.) to form the filler for the mix.

The limestone quarry for building stone is located on the north side of the site. It was from here that the limestone would have been sourced for burning in kilns to make the lime putty for the mortar.[8] It was not possible to burn lime on the site. The cathedral is located in a village and a licence to burn lime would not have been granted by the Environmental Protection Agency. In this instance it was necessary to buy in lime putty.

The burnt-clay component

There was definitely a burnt-clay component in the historic mortars examined. Was this deliberate? Could this small proportion have come from impurities in the local limestone? Some experimentation was required. Limestone was taken from the original quarry and sent to Narrow Water Lime Service in Warrenpoint, County Down, where Dan McPolin burnt it, and the result was analysed. The results showed the limestone was very pure with no clays present, so it became apparent that the clay must have been added in some way.

It is possible for some clay to have found its way into the kiln in small quantities before the burning took place. In the process of taking the quarried limestone to the kiln some clay could have adhered to the blocks. However, it is more likely that clay may have been used in the construction of the kilns and have become a very minor constituent in the firing process. It is interesting that analysis of other historic medieval mortars at the Office of Public Works sites also shows a small percentage of burnt clay. It is very curious that the amount has been fairly constant so far – about 1%. More research needs to be undertaken on this topic. It is also plausible that the medieval builders knew that the burnt clay gave the mortars a faster set and that its addition was deliberate.

Table C   Additives.

	Aggregate	Shells	Burnt clay	Charcoal	Ref. Sample analysis
Mix A1 11th C Stage 1 pointing	S1	5%	1%	2%	AC01
Mix A2 13th C Stage 1 pointing	S2	None	1%	2%	AC06
Mix A3 15th C Stage 1 pointing	S3	None	1%	None	AC09
Mix B 11th C mortar	S1	5%	1%	2%	AC01
Mix C 13th C mortar	S2	(5%)	1%	2%	AC06
Mix D 15th C mortar	S3	None	1%	None	AC09
Mix E 15th C Fine pointing	S8	None	1%	None	AC20
Mix F 13th C Fine pointing	S4	None	1%	2%	AC04
Mix G Tooled holes	S5	None	None	None	AC12
Mix H Putlog	S5	None	None	None	AC12
Mix I₁ 13th C plaster	S6	5%	1%	1%	AC12
Mix I₂ 15th C plaster	S7	2%	None	1%	AC14

Additives shall be added to dry aggregates only after aggregates have been thoroughly mixed. Clay to be added before use and this to be evenly mixed.

Note : Table B + Table C = Filler (100%)

Figure 5   Materials for mortars stored at the site. These were covered when not being used.

Mixing the replica mortars and plasters

All aggregates were sieved to achieve the correct range. They were stored at the site in one designated area on a hard, clean base and kept covered when not being used. The different materials were separated by partitions. One person was charged with the mixing to maintain consistency and good quality control.

The preparation of the materials was as follows:

- Prepare the secondary aggregates (SA1, SA2 etc.) so that they are thoroughly mixed and well graded.
- Prepare the primary aggregates by adding secondary aggregates in the specified percentages and mixing thoroughly to ensure even distribution of particles and a well-graded aggregate.
- Add additives to dry aggregate blends (PA1, PA2 etc.) to the percentages specified. Again thoroughly mix.
- Add filler (primary aggregate + additives) to the binder (lime) and mix in the ratios as set out in Table D. (The lime putty used was at least six to eight months old.)

Table D lists the final mixes according to the binder/filler ratio.

Table D   Final mix.

	Binder = Lime Putty	Filler = Aggregate + Additives
Mix A1 11th C Stage 1 pointing	1	3
Mix A2 13th C Stage 1 pointing	1	3
Mix A3 15th C Stage 1 pointing	1	3
Mix B 11th C mortar	1	3
Mix C 13th C mortar	1	3
Mix D 15th C mortar	1	3
Mix E 15th C Fine pointing	1	3
Mix F 13th C Fine pointing	1	3
Mix G Tooled holes	1	4
Mix H Putlog	1	4
Mix I$_1$ 13th C plaster	1	2½
Mix I$_2$ 15th C plaster	1	1½

Table E sets out the make-up of each of the mortar and plaster mixes. Examples of more detailed recipes have been given in the tables for Mixes B, C, D, I$_1$ and I$_2$ above.

Before any final decision was made on a mixer a sample mortar was mixed using three different kinds: a standard mixer, a mortar mill and a paddle mixer. The mortars were then examined and rated. The mortars mixed in the paddle mixer were by far the best quality. They were darker in colour, which suited the final finish, and they were exceptionally workable. The mix with the paddle mixer was very thorough and used less water; this may have resulted in the darker shade of mortar. As less water was required, when the mortar dried out and set it was less inclined to crack. So a paddle mixer was used throughout the project and the quality of the mixes was such that they are now used as standard at other sites.

Mortar samples were required before commencement of work to ensure correct colour, mix, performance and texture. Test cubes were sent for testing at regular intervals.

Table E   Mix details.

	BINDER	FILLER	Aggregates					Additives		
	Lime	Aggre-gates + Additives	A1 Quartz Sand	A2 Lime-stone Sand	A3 Lime-stone Gravel	A4 Sand-stone Sand	A5 Sand-stone Gravel	Shell	Burnt Clay	Char-coal
Mix A1 11th C Stage 1 pointing	100	300	270	None	6	None	None	15	3	6
Mix A2 13th C Stage 1 pointing	100	300	240	None	None	42	9	None	3	6
Mix A3 15th C Stage 1 pointing	100	300	129	129	3	30	6	None	3	None
Mix B 11th C mortar	100	300	270	None	6	None	None	15	3	6
Mix C 13th C mortar with shells	100	300	225	None	None	42	9	15	3	6
Mix D 15th C mortar	100	300	129	129	3	30	6	None	3	None
Mix E* 15th C Fine pointing	100	150	129	129	3	30	6	None	3	None
Mix F 13th C Fine pointing	100	300	141	75	None	75	None	None	3	6
Mix G Tooled holes	100	400	200	200	None	None	None	None	None	None
Mix H Putlog	100	400	200	200	None	None	None	None	None	None
Mix I₁ 13th C plaster	100	250	120	112½	None	None	None	12½	2½	2½
Mix I₂ 15th C plaster	100	150	30	7½	None	103½	4½	3	None	1½

Measurement 1 = 1 Unit

Figure 6   Experimenting with a mortar mill (top) and a paddle mixer (bottom) on the site prior to the work commencing.

WORK ON SITE

Pointing

Due to the long-term exposure of the walls inside and out, the original mortars had eroded considerably and one of the main conservation tasks was the repointing of almost the entire cathedral using the full range of replica mortars. There were vestiges of both thirteenth- and fifteenth-century internal plasters adhering to the walls, and the replica plasters were used just to repair the edges to help consolidate what little remained.

The walls were raked out, and degraded and decomposed mortar and later cementitious pointing repairs were removed. The open wall joints were then washed out under pressure to make sure that they were clean in order to take the new pointing. All original mortar which was in good condition and any sound plaster were left intact. The new mortars were applied with a full range of tools to make sure that they went well into the cavities between the stones, particularly in the decorative work with fine bedding joints. The mortar was finished off proud of the stone temporarily and left to initially set for approximately seven days (under good weather conditions); it was then finished off until it was flush with the stone. Finally, once it had hardened sufficiently, it was rubbed back to expose

Figure 7   Working at battlement level. The mortar is being well compacted into the wall and finished temporarily until the initial set (courtesy of Con Brogan, DAHG).

Figure 8   The south aisle after the mortar had been rubbed back to expose the aggregates and additives (courtesy of Con Brogan, DAHG).

Figure 9   Detail of the infill mortar at the top of the wall plaster rubbed back to expose the aggregates and additives.

the aggregates and additives. In time, the surface will be well washed and further exposed, giving the mortar its final visual and textural characteristics. This will help the visitor to the site to distinguish the different periods of construction.

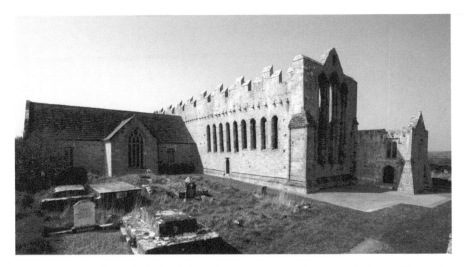

Figure 10   Ardfert Cathedral from the south-west after completion of Phase 2 (courtesy of Con Brogan, DAHG).

The work proceeded throughout the year and, though the cathedral is quite near the sea, there were times during the winter when frost was a problem. No work was undertaken once the temperature fell below 5°C. Hessian sacking was used to avoid damage due to the cold, extreme heat or excessively fast drying caused by the wind.

Given the size of the job there was a full-time clerk of works on the site and this helped in the maintenance of excellent standards and in dealing with problems as they arose. In addition, there was a full site meeting every two weeks so it was possible to set and maintain a good standard throughout the duration of the project.

Training of the masons

Training is a very important aspect of this type of specialist work. A budget was set aside in the contract to provide an element of training so that there could be better control of the standard of workmanship and finish. The subcontractor who provided the masons for the job was far-sighted enough to have all his men attend the training session. This was beneficial, as there were occasions when a mason might be off sick and a replacement had to be brought in from another site. In this case the mason had already been trained to work with these lime mortars.

CONCLUSION

We were very fortunate in this project to have had a reasonable budget for research and the time to undertake it. A great deal was learned in the process and this has helped define a template for other such projects. In time, it will be possible to build up a large body of information on historic mortars, which will form the basis for further research into the development and spread of historic mortars in Ireland.

As with every project there were a few glitches. Later in the project we had a problem with salts crystallizing on the surface of the mortars in two areas. Analytical work was undertaken to ascertain the cause of the problem, particularly to discover whether there was a problem with the lime-based grout used to fill the voids in the walls. The culprit, however, turned out to be the crushed shell. One batch had not been properly washed despite all precautions.

While there was a large research component in the use of the mortars, it is important to acknowledge that it is the quality of work on site that is important. It was interesting to see how positively the masons responded to working with lime mortars. Most had not used them before but, after initial caution and suspicion, they soon became quite expert in their use and impressed with the results they achieved. When works began at Listowel Castle in County Kerry shortly afterwards and we needed to recruit masons, some of those who had worked at Ardfert joined our direct labour workforce there. In doing so they had to take a cut in pay but they were willing to do this. Such is the power of lime mortar!

This paper was first published in the *Journal of the Building Limes Forum*, Volume 14, 2007. Note that there are small discrepancies in the tables between both papers. Those used here are the corrected versions.

Notes

1   This project was a joint project undertaken with Shaffrey Associates, who provided much of the back-up. It was groundbreaking on many fronts. Phase 2 of the works, which included this work on historic mortars, was undertaken jointly with Grainne Shaffrey. I would like to take this opportunity to thank both Grainne and her colleague, Eamonn Kehoe, who threw themselves so wholeheartedly into this project.

2   This is approximately £1¼ million sterling.

3   The analysis and reports on the mortar and plaster samples from Ardfert Cathedral were carried out and put together by Dr Sara Pavia on behalf of

Carrig, a specialist firm employing a range of conservation professionals. The information under this heading is abstracted from these reports.

4 For the purpose of this report the grading is: coarse (over 5 mm), medium (between 2 mm and 5 mm) and fine (less than 2 mm).

5 This work was carried out by Eamonn Kehoe of Shaffrey Associates.

6 The use of the word 'aggregate' can be confusing. In this paper its meaning includes both sands and aggregates.

7 This work was carried out locally by Eamonn Power MIEI, at the Institute for Technology in Tralee, County Kerry.

8 Dr Sara Pavia, who undertook the scientific analysis, does not consider that the original mortars were made using hot mixes.

Nano-Lime for Consolidation of Plaster and Stone

Paul D'Armada and Elizabeth Hirst

WHAT IS NANO-LIME?

Nano-lime consists of very small particles of calcium hydroxide suspended in alcohol; the average diameter is quoted as 150 nm with a range of 50–300 nm (Table 1).[1] Synthesized under specially formulated and controlled conditions, the production of nano-lime has no similarity with the slaking of lump lime to form lime putty. Nano-lime particles are much smaller than the conventional particles of calcium hydroxide present in fresh lime putties (8,000 nm) although research has shown that, in an aged lime putty, they can be as small as 200 nm.[2] The smaller particle size of nano-lime has the advantage of achieving greater penetration into the pores, while the higher surface area–volume ratio allows for greater reactivity.

Table 1 Sizes of different particles.

Particle	Ionic radius (nm)
Calcium ion	~ 0.1
Hydroxyl ion	0.14
Nano-lime	50–300
Conventional lime hydrate	Perhaps 10^3 to 10^5
Pore space	Perhaps 10^3 to 10^4

1 nm = 10^{-9} m
1000 nanometres (nm) = 1 micrometre (μm), also called a micron.
1000 micrometres = 1 millimetre

Although lime molecules (dissociated as calcium and hydroxyl ions) in limewater are even smaller (~0.24 nm) than nano-lime particles, water can dissolve only about 1.7 g of calcium hydroxide per litre at 20°C. So while lime water has traditionally been used as a consolidant on friable plaster and stone, it requires many saturations to get sufficient lime into the substrate and its consolidation effect can only be detected after an excess of 150–200 applications. Such copious amounts of water can actually cause deterioration of the stone or plaster through recrystallization of salts and other mechanisms of decay associated with repeated wetting and drying. As such the method may be impractical for a conservation project.[3]

The main theoretical advantages of nano-limes are:

- they can carry much greater quantities of lime because they are suspensions of lime (rather than solutions);
- no water is involved in the application.

Nano-limes satisfy the performance criteria usually expressed in any conservation strategy, that the materials used should, as far as is possible, be compatible with the material being conserved and be of predictable behaviour. But like most other consolidation techniques, nano-limes are not reversible. However, initial results on the efficacy and compatibility of nano-lime when used for the treatment of calcareous stone and plaster are thought to outweigh any disadvantage due to its irreversibility.

For the commercial formulations used in tests and in the project at All Saints' Church, Little Kimble, described below, the quantity of lime carried is reflected in the name of the product. For example, CaLoSiL® E-50 carries 50 g of lime per litre, or approximately 30 times more lime than in limewater. Compared with limewater, this particular nano-lime product could theoretically achieve equivalent deposition of calcium hydroxide into a porous material with only a thirtieth the number of applications.

HOW DOES NANO-LIME WORK?

When a substrate is treated with nano-lime, calcium hydroxide is precipitated in the pores of calcareous materials as the alcohol disperses and/ or evaporates. As calcium hydroxide carbonates to calcium carbonate, it replaces lost binder or matrix in natural stone and plasters, thereby 'knitting' together fine cracks and deteriorated stone, and increasing strength and integrity.

The overall carbonation process is represented by the following reaction:

$$Ca(OH)_2 \quad + \quad (H_2O + CO_2) \quad \rightarrow \quad CaCO_3 \quad + \quad 2H_2O$$

calcium   carbonic   calcium   water

hydroxide (lime)   acid   carbonate

Both water and carbon dioxide are necessary for carbonation of the lime, but the carbon dioxide can only react with lime by intimate molecular contact when it is carried in water as carbonic acid (H_2CO_3). The water required can come from the atmosphere, the treated material, spray application or by adding it to the nano-lime before application. When water is added to nano-lime to increase its reactivity, however, the resultant formation of a calcium hydroxide gel may reduce penetration. Like any slaked lime, nano-lime carbonates more slowly at lower temperatures and will not carbonate at all at 4°C or below.

Consolidants based on the carbonation of calcium hydroxide may not be appropriate for the treatment of pure sandstones because they may introduce calcium salts such as calcium sulfate (gypsum), with well-catalogued detrimental results.

APPLICATION OF NANO-LIME

Current research suggests that nano-limes have a shelf life of 3–5 months (longer if refrigerated). Suspensions are available in ethanol (the 'E' series), isopropanol and n-propanol; those in ethanol have a slightly lower viscosity. Nano-lime can be applied by brushing, injection, spray, pouring, immersion, vacuum impregnation and systematic dripping techniques. Brushing is not as effective as the other methods and poulticing appears to be generally ineffective, particularly when the nano-lime has a greater affinity with the poultice material (e.g. paper) than with the minerals in the substrates.

For any decayed material, it is important that the degraded zone is saturated with consolidant to a depth beyond the sound material. Treatment can be repeated when all the solvent has evaporated and carbonation of the nano-lime has occurred. The depth to which nano-lime will penetrate into a porous material depends largely on the character and openness of the surface treated, and the prevailing environmental conditions. For example, compact surfaces and those with a gypsum or mineralized crust usually prevent satisfactory penetration while damp, cold conditions will inhibit evaporation of the alcohol.

Capillary theory suggests that the rate of diffusion of the nano-lime is 2–3 times slower than its alcohol dispersant which, in turn, is approximately 2–3 times slower than water (or limewater). The type of consolidant used and its surface tension, viscosity and volatility are, therefore, important factors, not only for the speed of impregnation but for the quantity of consolidant used (some of it being lost by evaporation). For example, because alcohol not only diffuses more slowly than water but evaporates more quickly, it is more difficult than water to introduce into a porous material. But in cracks and fissures, and where the pores are larger and have less capillarity, alcohol will generally flow better than water. Being non-polar, its molecules are less cohesive than water and, as a result, alcohol wets better than water.

Nano-lime can be used on its own for the consolidation and strengthening of stone, plasters and mortars, or in combination with aggregates and fillers to produce injection grouts or repair mortars.

BACKGROUND TO RESEARCH ON NANO-LIME

The production and utilization of nano-sized suspensions of lime has been studied for more than a decade. The first use of reasonably stable lime dispersions in n-propanol was reported in 2000 by Giorgi, Dei and Baglioni.[4] They studied capillary suction and water vapour permeability of treated and untreated specimens of lime sand mortars, and followed the carbonation of the nano-lime by X-ray diffraction (XRD). The formulations prepared by this team were used for the treatment of wall paintings in Santa Maria Novella in Florence, the nano-lime being applied by brush through Japanese tissue.

Dei subsequently shifted his research to isopropanol dispersions. His work with nano-lime on wall paintings at the Museo del Bargello was published in 2005, with particular attention being given to the question of colour stability.[5] In the following year, he co-authored a scanning electron microscopy (SEM)/energy dispersive X-Ray spectroscopy (EDX) investigation of the mechanism of consolidation, which also discussed the treatment of limestone.[6,7] A 2007 paper presented some additional details on dilutions and multiple applications carried out in conjunction with the treatment of a wall painting by Agnolo Gaddi in the church of Santa Croce (Florence).[8] The other two members of the 2000 team (Baglioni and Giorgi) extended their work with nano-lime beyond Europe, using it to stabilize wall paintings (and limestone) at the ancient Mayan city of Calakmul in Campeche, Mexico.[9]

The STONECORE project

In September 2008 the STONECORE project ('Stone Conservation for the Refurbishment of Buildings') was established with funding from the 7th Framework Programme for Research (FP7) of the European Commission. Coordinated by IBZ-Freiburg (a German research and development company that has now made nano-limes commercially available to conservators), it is a consortium of researchers representing several private companies and public institutions, including the Technical University Delft and the Czech Academy of Science.

Research and tests carried out at part of the STONECORE project included in situ drill resistance, ultrasound transmission and ground-penetrating radar measurements, as well as more conventional laboratory tests for gains in compressive and flexural strength.[10] Although they were presented at conferences on stone conservation, two papers reported on the testing of nano-lime on specimens of lime mortars.[11,12]

Most of the STONECORE findings were disseminated to the conservation community through a series of public meetings. These were held at Litomysl (Czech Republic) and Torun (Poland) in 2010 and at Peterborough (UK) and Freiberg (Germany) in 2011. Speakers at the last of these conferences were from Austria, the Czech Republic, Germany, Greece, the Netherlands, Poland and the UK.

The STONECORE project ended in August 2011. Additional information is available from the project website (www.stonecore-europe.eu), which includes links to the partners.

SOME FINDINGS FROM THE STONECORE PROGRAMME

Although many measurements related to laboratory experiments and much of the site work focused on mortars and plasters, the papers prepared as part of the STONECORE programme showed that there is real potential for the use of nano-lime as a consolidant for calcareous substrates, as well as having other possible uses such as a biocide. Some of the results are summarized below.

Despite reductions in porosity of up to 5% after consolidation with nano-lime, it was found that the treatment has little influence on the water transport properties (capillary suction) of the treated materials. Indeed, it can improve these properties where it bridges cracks and allows coherent diffusion of moisture.

Ultrasonic velocity and drill resistance measurements showed that:

- the compressive, flexural and surface cohesion strengths of materials treated with nano-lime increase with the number of applications;
- compressive strength generally increases more rapidly than flexural strength as the number of applications or saturations with nano-lime increases.

Nano-lime in ethanol seems to be a more effective, general purpose consolidant than nano-lime in isopropanol. Test results clearly show that CaLoSiL® E-25 gives higher increases in compressive strength and tensile strength than CaLoSiL® IP-25 (or E-50) for consolidation of limestones and crushed stone dusts (Table 2). The percentage increase in tensile strength, however, was less with E-25 than IP-25 because the isopropanol dispersant did not penetrate as well as ethanol; the nano-lime in the former, therefore, tends to reside in greater concentrations on or near the surface of the treated material. It has been demonstrated that when a consolidant diffuses more uniformly throughout a material, optimum gains in compressive strength are achieved, whereas if more of it resides or precipitates closer to the surface, tensile strength increases at the expense of the compressive strength.

Table 2   Relative increase in compressive strength of Maastricht limestone using CaLoSiL® E-25 and IP-25.

Number of saturations	% increase in compressive strength	
	E-25	IP-25
2	50	23
4	53	43
6	93	47

There have been a number of reports of white haze formation from the carbonation of lime on the surface. This can be mitigated by either starting with a low concentration of nano-lime and increasing it with each application, or preventing evaporation of the alcohol (which might otherwise cause feedback of the nano-lime to the surface) and removing any excess nano-lime immediately after treatment with a sponge. White haze formation on the surface is more likely when using nano-lime dispersed in isopropanol, because having a higher molecular weight than ethanol, it does not penetrate as quickly. If necessary, the white lime haze can be removed later using steam.

Results also suggested that the denser the stone or the more compact the aggregate in a plaster or mortar, the higher the consolidation strength achieved.

Nano-lime used in combination with ethyl silicates

Tests carried out as part of the STONECORE project suggest that nano-lime may also be used in conjunction with ethyl silicates, especially when consolidating larger voids and areas of delamination, even in very damp outdoor conditions. This is because nano-lime acts as a catalyst for the hydrolysis of the silicic acid ester (ethyl silicate) and as a coupling agent, enhancing the bond to stone surfaces. The amorphous calcium silicate hydrate gel formed can fill larger voids than silicic acid ester or nano-lime alone, and becomes hydrophilic faster than ethyl silicate; however, the water absorption and capillarity of the treated material may be lowered as a result. Tests and assessments using SEM demonstrate that the strengths of degraded materials can be doubled with only one pre-treatment with nano-lime followed by one treatment with an ethyl silicate.

Further research into the combined use of nano-lime and ethyl silicates on possible adverse effects with salts such as gypsum is necessary before this treatment can be fully approved. This is particularly important for the conservation of historic fabric, which is often persistently damp and therefore likely to contain salts (e.g. sulphates, chlorides and nitrates), which have probably contributed to the failure in the first place.

Project-oriented studies

The treatment carried out to the wall paintings at All Saints' Church in Little Kimble, Buckinghamshire, in 2010 was the first large-scale use of nano-lime in the UK. The decision to use it was partially guided by the in-house programme of laboratory studies summarized below. Some related studies were performed by Emily Howe in 2006–07 in conjunction with the treatment (with E-25) of small areas of the Lichfield Angel, an early medieval limestone sculpture excavated in 2003.[13] Equally encouraging were the many Italian nano-lime publications on mural treatment and the STONECORE presentations by Musiela describing the work on the cellars of the Middle

Castle, Malbork, on the façade of the Cathedral of the Visitation Order in Warsaw, and in the Cathedral Basilica in Torun in Poland.[14]

Diffusion of nano-lime in ethanol (CaLoSiL® E-25)

As part of the in-house experiments, the transport characteristics (by capillarity) and penetration rates of E-25 into laboratory samples of three unweathered UK limestones (Weldon, Ketton and Clipsham) were measured using a continuous feed of E-25 into the stone surface and negligible loss of ethanol by evaporation (Plate 32). Because evaporation of the ethanol on the surfaces of the material to be treated would considerably reduce diffusion rates, the rates reported here may be the maximum possible without vacuum impregnation. The data are shown in Table 3 and presented graphically in Figure 1.

Table 3 Rates and distance of capillary diffusion in three different limestones (Hirst Conservation).

Limestone type	Rate of diffusion of CaLoSiL® E-25 (cm per minute)	Rate of diffusion of water (cm per minute)	Water absorption % by weight (Stone Directory)	Water absorption % by weight (test sample)	Porosity % by Volume (test sample)	Density g/cm3 (test samples)	Distance of penetration of nano-lime (cm)
Weldon	~ 0.45	~ 1.4	–	8.5	18.4	2.0	5.5
Ketton	~ 0.1	~ 0.4	9.8	6.7	15.4	2.14	4.5
Clipsham	~ 0.02	~ 0.08	4.7	7.7	17.5	2.1	4.2

Rates of diffusion of E-25 are from 0.02 to 0.5 cm per minute. The variation is considerable from stone to stone, despite only modest differences in measured water absorption and porosity. This is presumably because of differences in pore size distribution. The water movement data parallel the results for E-25; water penetrates considerably faster than E-25, primarily because water has a surface tension more than three times that of ethanol.

The maximum depth of deposition of the nano-lime particles, as measured with phenolphthalein, was 4–5.5 cm (consistent with results from other researchers). For most consolidation of plaster and calcareous stone,

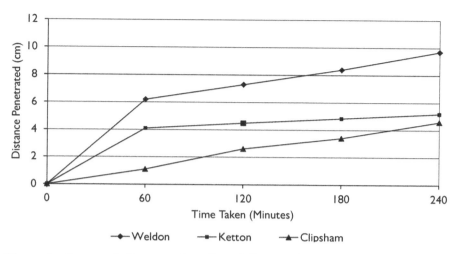

Figure 1   Rates and distance of capillary diffusion of CaLoSiL® E-25 in three different limestones, with inhibited evaporation of its ethanol dispersant (Hirst Conservation).

Figure 2   Cube consolidated to a depth of 1.5 cm by 20 saturations with nano-lime (E-25).

this depth of treatment is satisfactory. While the test samples had not deteriorated, weathered stone would be expected to show even deeper and faster transport of the nano-lime. In a second round of phenolphthalein testing, carbonation of the material deposited in the pores was observed after three days (Figure 2).

Consolidation of powdering paint

Limited laboratory tests were undertaken to determine whether unbound (pigments applied in water) paint surfaces could be sufficiently consolidated with nano-lime when applied in increasing saturations and different concentrations.

Panels of plaster board were painted with red ochre pigment dispersed in water so that the resultant paint film would be completely unbound upon drying. The boards were then treated with the nano-lime solutions and left uncovered to allow carbonation of the consolidant.

When lightly abraded with a finger, none of the paint was significantly stabilized or consolidated. Scotch tape further demonstrated the lack of pigment cohesion. The results of these preliminary trials showed that the nano-lime tended to carbonate on the surface rather than within the pigmented layer, which remained decohered. The higher concentrations showed more of a white lime haze on the surface of the paint.

The reported successes in the treatment of plaster and wall paintings, including frescoes, strongly suggest that further testing is required – perhaps using pigments bound in different media, followed by further appraisal and more specific tape testing.[15]

RECENT USE OF NANO-LIME AT ALL SAINTS' CHURCH, LITTLE KIMBLE

Introduction

Following the successful use of CaLoSiL® products in several European projects, its use was considered by Hirst Conservation for the consolidation of a scheme of historically important medieval wall paintings at All Saints' Church, Little Kimble, Buckinghamshire.

Little Kimble is a village at the foot of the Chiltern Hills, about 5 km south of Aylesbury. The Grade I listed church, part of the Diocese of Oxford, is a small, flint-clad building constructed in the twelfth century. The walls are decorated with a highly significant scheme of wall paintings, dated to the early fourteenth century.[16] Today, the most important surviving paintings, found principally on the nave walls, are the scenes of saints and of the Virgin and Child.[17] Perhaps the best known of these is a depiction on the north wall of St George as a Knight Templar, bearing a shield with the red-on-white cross.

Condition

The wall paintings were treated by Professor E. W. Tristram, *c.* 1930. The surfaces were covered with wax, which was a favoured treatment at that time as it visually strengthened the image and consolidated the paint coating. Unfortunately, this type of impervious coating resulted in problems with many British wall paintings, as the movement of moisture and salts inevitably caused failure of these early paintings especially in damp English churches. Eve Baker and her team removed this wax from many medieval paintings and work was also done at Little Kimble, *c.* 1972. Professor Robert Baker and Mrs Baker also secured the surfaces with sensitive repairs and grouting, using non-hydraulic lime plaster. Their restoration has proved to be effective as the condition of the wall paintings at Little Kimble has remained generally stable, despite poor environmental and building conditions.

In recent years the parish has endeavoured to improve the overall condition of the building. This included a programme of roof repairs completed in 2004 and further building works undertaken in 2007. Improved drainage was installed in 2009 to reduce rising damp.

A condition survey was undertaken in 2005 when the entire surface of each of the wall paintings was gently tapped to detect areas of instability (possible delamination of the plaster from the stone). In many instances, the hollows detected were stable and corresponded with areas of old losses and cracks that had since been repaired. However, some of these areas displayed movement when gently pressed, indicating the long-term potential for loss and the need for repair, although comparison with an earlier condition report in 2001 suggested that the wall paintings were relatively stable.[18]

The investigations included monitoring of the environmental conditions within the church, and an assessment of the impact of the building envelope on the condition of the medieval wall paintings. The church had a hard plaster dado, which was applied as an earlier intervention prior to the Baker restoration. The impermeable nature of the plaster would have caused rising damp to be forced higher up the wall and into the region of the paintings. It was considered beneficial to remove this plaster and replace it with a breathable lime plaster.

On close inspection, original surfaces were powdering and veiled with cobwebs, bat guano, dirt and salts. Salt pustules had erupted on the paintings in many places and the edges of paint films had delaminated in areas adjacent to some of the repairs. Numerous hairline cracks in the surface could also be observed. Some previously repaired and filled cracks, most notably in the window reveals and around the doorways, had begun to open up again.

It was apparent that, while the paintings themselves were quite stable, the walls had many later plaster repairs that had begun to fail, which was particularly obvious at lower levels, with patches of white efflorescence and blistering. The previous sensitive, tinted lime repairs and localized limewashing around the paintings done by Eve Baker remained sound.

As lime was a successfully 'tried and tested' material in this church, it supported the rationale that, given the ongoing problems of moisture movement, intervention should be based on porous materials and lime-based treatments. Accordingly, the use of nano-lime on the wall paintings was fully justified for this project. As a stable environment within the church is not obtainable in the foreseeable future, it seemed inappropriate to make extensive use of synthetic resins. The nano-lime suspension, CaLoSiL®, was chosen as a consolidant because it is believed to allow for better evaporation of moisture from the damp fabric within the church.[19]

A subsequent re-examination was performed in June 2009 to update the condition report in anticipation of the conservation treatment.[20] The treatment of the paintings was discussed with the client[21] and English Heritage[22] before final specifications were agreed and works instructed.

Conservation

The conservation work was undertaken in 2010. Emergency consolidation with Plextol B500 acrylic dispersion was required in a few localized areas but, in general, avoided in favour of nano-lime. Bat and bird droppings, cobwebs, dust, loose and imbibed dirt, together with some salt efflores-cence, were removed from the surface.

Failing, cementitious and gypsum plasters of no historic importance, mostly at dado level below the historic paintings, were taken off. These were replaced with lime plasters to improve the aesthetics of the church interior and to allow moisture in the walls to evaporate at heights lower than the paintings.

Larger delaminations, cracks and voids could be stabilized using non-hydraulic lime putty and fine aggregate (sand and stone dust). Nano-lime materials were used for the fine cracks and fissures to avoid the introduction of undesirable amounts of water into the plaster and the drilling of holes to facilitate the use of catheter tubes and pipettes. Most of the cracks in the wall plaster at Little Kimble were hairline or very fine, and not particularly unsightly or disfiguring. Such cracks are ordinarily left unfilled as potential stress relief opportunities if the wall structures should move due to thermal or hygric fluctuations or building subsidence. The challenge in this instance was that they were often adjacent to delaminating plaster and paint layers that required stabilization. Plate 33 shows the areas of repair and repair types to the wall painting of St George.

As the walls might have contained some residual moisture in places, it was deemed beneficial to pre-inject the relevant areas or cracks with industrial methylated spirits (IMS) to help dry them out (Figure 3). This would allow better penetration of the nano-lime as moisture-filled pores impede diffusion. CaLoSiL® E-25 and/or E-50 were injected slowly into the cracks until apparent saturation of the adjacent deteriorated plaster had occurred. The product was not allowed to bleed onto the surface of the painting, thereby avoiding formation of the unsightly haze of lime that can occur as alcohol (in this case ethanol) evaporates. (If it does occur, it can be removed quickly with cotton wool dampened in IMS.)

It can take up to 24 hours for the alcohol in nano-lime formulations to evaporate. The result is deposition of the nano-lime particles in the pores of the plaster and between fine layers of delamination. In this project, 8–10

Figure 3   Detail showing the pre-wetting of an area to be consolidated using IMS.

Figure 4 Consolidation of the paint layers of the wall painting depicting St George (left) and consolidation of the wall-painting depicting St Clare using CaLoSiL® E-25 (right).

applications to saturation were applied by syringe over several days to all deteriorated areas. On average, ten 10–50 ml applications of E-25 and/or E-50 per application were applied in each area requiring consolidation.

Because carbonation occurs more quickly near the surface, better consolidation might have been achieved by allowing each application of nano-lime to carbonate fully before the next was applied. However, this could have taken days or weeks; the limitations of a site-based project did not allow more than 24 hours between applications. Some areas were consolidated using this method before injecting lime-based grouts. In addition, some larger fragments and areas of delamination were treated with CaLoXiL® injection grout, which is a more concentrated form of nano-lime in ethanol also containing marble powder.

As described above, tests made before the project began both informed and guided the decision to use the nano-lime method. Although no quantitative evaluation was undertaken in situ after treatment, a gentle tapping ('sounding') after several days of the formerly unstable areas of plaster indicated that these were now well-adhered and stable. Moreover, in areas where the nano-lime had been injected or carefully brushed behind loose flakes of thin paint and/or plaster layers, the consolidation was also successful without any visible change to the surface. Figure 4 shows work in progress on the wall paintings of St George and St Clare.[23]

FOR THE FUTURE

There is little doubt that nano-lime is an important addition to the range of materials suitable for consolidation of porous substrates. But despite the rapid increase in the literature on nano-lime as a conservation material, it is clear that more technical and practical studies need to be undertaken. For practitioners, it is important to have a broader range of published case studies that describe the materials and conditions most suited to nano-lime treatment.

One key issue is effectiveness. This is often judged in the laboratory on specimens that do not fully simulate the circumstances encountered in the field. Where field testing has been done, in situ evaluation is frequently undertaken too soon after treatment, that is, without taking account of the aging effects of fluctuating temperature and moisture, and the cyclical crystallization of salts.

To provide an enhanced picture of the potential value of nano-limes in the consolidation of plaster, there should be more comparative field testing and sharing of findings through conferences and publications. Those significant full-scale treatments that have been carried out in the past ten years should be revisited and the results re-evaluated to improve the way in which nano-limes are used in the next decade.

This would be of enormous value to the conservation profession, aiding in decision-making when the details of the treatment (product, diluent, concentration, application technique) are supplemented by environmental monitoring data. As part of this process, the Building Conservation Research team of English Heritage has begun a two-year research project to provide greater understanding of the parameters for site-based treatment of weathered limestones. This project is being carried out in conjunction with the Department of Architecture and Civil Engineering at the University of Bath.

For plates referred to in this paper, see the colour section following page 132.

This paper was first published in the *Journal of Architectural Conservation*, Volume 18, No. 1, March 2012.

Acknowledgments

The authors would like to thank Norman Weiss and Kyle Normandin for their contribution to this paper regarding the international studies into nano-lime, and David Odgers for his valuable editorial support as well as Ian Brocklebank and the Hirst team. Gerald Zeigenbalg and the STONECORE team supported this research and John Fothergill of Kimble Parochial Church Council gave endless encouragement for the work at Little Kimble.

Notes

1 Ziegenbalg, G., Bruemmer, K. and Pianski, J., 'Nano-lime – a new material for the consolidation and conservation of historic mortars', *2nd Historic Mortars Conference*, RILEM, Prague, 2010, pp. 1301–9.

2 Cazalla, O., Rodriguez-Navarro, C., Sebastian, E. and Cultrone, G., 'Aging of lime putty: Effects on traditional lime mortar carbonation', *Journal of the American Ceramic Society*, Vol. 83, No. 5, 2000, pp. 1070–6.

3 For a summary of the literature on limewater see: Hansen, E., Doehne, E., Fidler, J., Larson, J., Martin, B., Matteini, M., Rodriguez-Navarro, C., Pardo, E. S., Price, C., de Tagle, A., Teutonico, J. M. and Weiss, N., 'A review of selected inorganic consolidants and protective treatments for porous calcareous materials', *Reviews in Conservation*, Vol. 4, 2003, pp. 13–25.

4 Giorgio, R., Dei, L. and Baglioni, P., 'A new method for consolidating wall paintings based on dispersions of lime in alcohol', *Studies in Conservation*, Vol. 45, No. 3, 2000, pp. 154–61. See also: Giorgio, R., Dei, L. and Baglioni, P., 'Nuevo métodopara la consolidación de pinturas murales basado en las dispersiones de cal viva en alcohol' [A new method for mural painting consolidation, based on calcium oxide/alcohol dispersions], *Revista PH*, Vol. 9, No. 34, 2001, pp. 57–63.

5 Dei, L., Radicati, B. and Salvadori, B., 'Sperimentazione di unconsolidante a base di idrossido di calcionanofasicosugliaffreschidella Cappella del Podestà al Museo del Bargello di Firenze: aspettichimico-fisici e prove di colore' [Study on a consolidation treatment based on nano-phase calcium hydroxide ...], *Sulle Pitture Murali*, Arcadia Ricerche, Bressanone, 2005, pp. 99–108.

6 Dei, L. and Salvadori, B., 'Nanotechnology in cultural heritage conservation: nanometric slaked lime saves architectonic and artistic surfaces from decay', *Journal of Cultural Heritage*, Vol. 7, No. 2, 2006, pp. 110–15.

7 Daniele, V., and Taglieri, G., 'Nanolime suspensions applied on natural lithotypes: the influence of concentration and residual water content on carbonation process and on treatment effectiveness', *Journal of Cultural Heritage*, Vol. 11, No. 2, 2010, pp. 102–6.

8 Dei, L., Bandini, F., Felici, A., Lanfranchi, M., Lanterna, G., Macherelli, A. and Salvadori, B., 'Preconsolidation of pictorial layers in frescoes', *Il Consolidamento degli Apparati Architettonici e Decorativi*, Arcadia Ricerche, Bressanone, 2007, pp. 217–24.

9 There are several publications on the work in Mexico. See especially:

Baglioni, P., Vargas, R. C., Chelazzi, D., González, M. C. G., Desprat, A. and Giorgi, R., 'The Maya site of Calakmul: *In situ* preservation of wall paintings and limestones using nanotechnologies', *The Object in Context*, Munich Congress, London, 2006, pp. 162–9.

10 Dimitriadis, K., 'The ground penetrating radar method used as a tool for monument restoration: Proposed usage and limitations in the detection of stone damage', presented at the Stonecore conference, 'Stone Conservation for refurbishment of Buildings', 23 August 2011, Freiberg, Germany.

11 Drdacky, M., and Slizkova, Z., 'Calcium hydroxide based consolidation of lime mortars and stone', *Stone Consolidation in Cultural Heritage*, LNEC, Lisbon, 2008, pp. 299–308.

12 Ziegenbalg, G, 'Colloidal calcium hydroxide: a new material for consolidation and conservation of carbonatic stones', 11th International Congress on Deterioration and Conservation of Stone, Torun, Poland, 2008, pp. 1109–15.

13 Howe, E., *A report on the conservation of the Lichfield Angel*, report dated May 2007.

14 See www.restauro.pl for information on these nano-lime projects.

15 Tape testing of surfaces treated with nano-lime was discussed by Giorgi, Dei and Baglioni in their 2000 article (see note 4). It is a technique often used for the study of paint chalking. See Methods C and D of ASTM D4214-07 Standard test methods for evaluating the degree of chalking of exterior paint films.

16 Pevsner, N., *The Buildings of England: Buckinghamshire*, Penguin Books, Harmondsworth, reprinted 1979, p. 314.

17 Tristram, E. W., *English Wall Painting of the Fourteenth Century*, Routledge & Paul, London, 1955, pp. 188–9.

18 Lithgow, R., *Revised preliminary survey of wall paintings, All Saint's Church, Little Kimble,* report dated 25 January 2001.

19 The manufacturer's website is www.ibz-freiberg.de.

20 Hirst Conservation, *Condition survey of the fourteenth century wall paintings in All Saints Church, Little Kimble, Buckinghamshire*, revised report dated August 2009.

21 Andrew Argyrakis of the Church Buildings Council (CBC), formerly the Council for the Care of Churches (CCC).

22 Represented by Robert Gowing.

23 See Tristram, E. W., *op. cit.* for a description of these two wall paintings.

Further Reading

Allen, G., Allen, J., Elton, N., Farey, M., Holmes, S., Livesey, P., and Radonjic, M., *Hydraulic Lime Mortar for Stone, Brick and Block Masonry*, Donhead Publishing Ltd, Shaftesbury, 2003.

American Society for Testing Materials, *ASTM C1324-10, Standard test method for examination and analysis of hardened masonry mortar*, ASTM International, Philadelphia, USA, 2010.

Anon, *Practical Remarks on Cements for the Use of Civil Engineers, Architects, Builders etc*, second edition, William Gilbert, London, 1832.

Arcolao, C., *Le ricette del Restauro. Malte, intonaci, stucchi dal XV al XIX secolo*, Saggi Marsilio, Venezia, 1998.

Ashall, G., Butlin, R. N., Teutonico, J. M. and Martin, W., 'Development of lime mortar formulations for use in historic buildings', *Durability of Building Materials and Components 7 – Proceedings of the Seventh International Conference on Durability of Building Materials and Components*, ed. Sjöström, C., Stockholm, 19–23 May, E. & F. N. Spon, London, 1966.

Ashurst, J., *Mortars, Plasters and Renders in Conservation* (2nd edition), Ecclesiastical Architects and Surveyors Association, London, 2002.

Ashurst, J. and Ashurst, N., *Practical Building Conservation, Volume 3: Mortars, Plasters and Renders*, English Heritage Technical Handbook Series, Gower Technical Press, Aldershot, 1988.

Ashurst, J. and Dimes, F. G., *Conservation of Building and Decorative Stone*, Butterworth Heinemann, Oxford, 1998.

Banfill, P. F. G. and Forster, A. M., 'A relationship between hydraulicity and permeability of hydraulic lime', *Proceedings of RILEM International Workshop PRO12*, Paisley, 2000.

Barnes, P., *Structure and Performance of Cements*, Applied Science Publishers, London, 1983.

Bayer, K., et al., 'Microstructure of historic and modern Roman Cements to understand their specific properties', *13th Euroseminar on Microscopy Applied to Building Materials*, Ljubljana, Slovenia, 14–18 June 2011.

Bergoin, P., tr. Spano, U., 'Lime after Vicat', *Lime News*, Vol. 7, 1997.

Bessy, G. E., 'The maintenance and repair of Regency painted stucco finishes', *RIBA Journal*, Vol. 57, 1950.

Biston, V., *Manuel théorique et pratique du chaufournier*, Libr. encycl. de Roret, Paris, 1836.

Blezard, R., 'Chapter 1. History of calcareous cements', *Lea's Chemistry of Cement and Concrete*, Fourth Edition, ed. Hewlett, P. C., Arnold, London, 1998.

Boero, J., *Fabrication et emploi des Chaux hydrauliques et des ciments*, Librairie Polytechnique Ch. Beranger, Paris, 1901.

Boynton, R. S., *Chemistry and Technology of Lime and Limestone*, 2nd edition, John Wiley & Son, New York, 1980.

Brandon, C., 'Caesarea Papers 2 – Pozzolana, lime and single-mission barges (Area K)', *Journal of Roman Archaeology*, Supplementary Series Number 35, 1999.

BRE, *BRE DIGEST 360. Testing bond strength of masonry*, BRE, Garston, April 1991.

BRE, *BRE Digest 421. Measuring the compressive strength of masonry materials: the screw pull-out test*, BRE, Watford, 1997.

Bristow, I. C., 'Exterior renders designed to imitate stone: A review', *Annual Transaction*, Association for the Studies in the Conservation of Historic Buildings, Vol. 12, 1997.

British Standards Institution (BSI), *BS 890:1940. Specification for Building Limes*, BSI, London, 1940.

British Standards Institution (BSI), *BS CP 111. Structural recommendations for loadbearing walls*, BSI, London, 1948, 1964 and 1970.

British Standards Institution (BSI), *Code of Practice 121.201, Masonry, Walls – Ashlared with Natural Stone or with Cast Stone*, British Standards Institution, London, 1951.

British Standards Institution (BSI), *BS 890:1966. Specification for Building Limes*, BSI, London, 1966.

British Standards Institution (BSI), *EN 1015-11. Methods of test for mortar for masonry: Determination of flexural and compressive strength of hardened mortar*, BSI, London, 1999.

British Standards Institution (BSI), *EN 1052-1. Methods of test for masonry: Determination of compressive strength*, BSI, London, 1999.

British Standards Institution (BSI), *EN 1052-2. Methods of test for masonry: Determination of flexural strength*, BSI, London, 1999.

British Standards Institution (BSI), *BS 890:1972. Specification for Building Limes*, BSI, London, 1972.

British Standards Institution (BSI), *BS 890:1995. Specification for Building Limes*, BSI, London, 1995.

British Standards Institution (BSI), *BS EN 8221-2. Code of practice for cleaning and surface repair of buildings – Part 2: Surface repair of natural stones, brick and terracotta*, BSI, London, 2000.

British Standards Institution (BSI), *BS EN 459:2001. Building Lime*, BSI, London, 2001.

British Standards Institution (BSI), *EN 1052-3. Methods of test for masonry: Determination of initial shear strength*, BSI, London, 2002.

British Standards Institution (BSI), *British Standard 5628-1:2005. Code of practice for the use of masonry. Structural use of unreinforced masonry*, BSI, London, 2005.

British Standards Institution (BSI), *EN 1052-5. Methods of test for masonry: Determination of bond strength by the bond wrench method*, BSI, London 2005.

British Standards Institution (BSI), *BSI 4551. Mortar – Methods of test for mortar. Chemical analysis and physical testing*, BSI, London, 2005

British Standards Institution (BSI), *EN 1996-1. Eurocode 6. Design of masonry structures: General rules for reinforced and unreinforced masonry structures*, BSI, London, 2005.

British Standards Institution (BSI), *BS EN 13914-2. Design, preparation and application of external rendering and internal plastering – Part 2: Design considerations and essential principles for internal plastering*, BSI, London, 2005.

British Standards Institution (BSI), *BS 4551-2005 Mortar. Methods of test for mortar. Chemical analysis and physical testing*, BSI, London, 2010.

British Standards Institution (BSI), *EN 459-1:2010. Building Lime Part 1: Definitions, specifications and conformity criteria*, BSI, London, 2010.

Burn, R. S., *The New Guide to Masonry, Bricklaying and Plastering*, John G. Murdoch, London, 1871, reprinted by Donhead Publishing Ltd, Shaftesbury, 2001.

Burnell, G. R., *Rudimentary Treatise on Limes, Cements, Mortars, Concretes, Mastics, Plastering etc*, 6th edition, Virtue & Co., London, 1867.

Cameron, D. G., *Directory of Mines and Quarries*, British Geological Survey, Keyworth, 2008.

Candlot, E., *Ciments et Chaux Hydrauliques*, Ch. Béranger, Paris, 1906.

Caxton House Editorial, *Specification*, incorporating the Municipal Engineer's Specification, No. 11, London, 1908–9.

Cazalla, O., Rodriguez-Navarro, C., Sebastian, E. and Cultrone, G., 'Aging of lime putty: Effects on traditional lime mortar carbonation', *Journal of the American Ceramic Society*, Vol. 83, No. 5, 2000.

Chateau, T., *Technologie du Batiment*, Libraire d'Architecture de B. Bance, Paris, 1863.

Collepardi, M., 'Thaumasite formation and deterioration in historic buildings', *Cement and Concrete Composites*, Vol. 21, No.2, 1999.

Corish, A and Jackson, P., 'Portland cement properties – past and present', *Concrete*, July 1982.

Corish, A., 'Portland cement properties – updated', *Concrete*, January/February 1994.

Cowper, A. D., *Lime and Lime Mortars*, BRE, London, 1927, reprinted by Donhead Publishing Ltd, Shaftesbury, 1998.

Daniele, V., and Taglieri, G., 'Nanolime suspensions applied on natural lithotypes: The influence of concentration and residual water` content on carbonation process and on treatment effectiveness', *Journal of Cultural Heritage*, Vol. 11, No. 2, 2010.

Davey, N., *A History of Building Materials*, Phoenix House, London, 1961.

Dearn, T. D. W., *The bricklayer's guide to the mensuration of all sorts of brick-work, … with observations on the causes and cures of smoky chimnies, the formation of drains, and the best construction of ovens*, J. Taylor, Architectural Library, London, 1809.

Deer, W. A., Howie, R. A., and Zussman, J., *An Introduction to the Rock Forming Minerals*, Longman Publications, London, 1989.

Dei, L. and Salvadori, B., 'Nanotechnology in cultural heritage conservation: nanometric slaked lime saves architectonic and artistic surfaces from decay', *Journal of Cultural Heritage*, Vol. 7, No. 2, 2006.

Dei, L., Bandini, F., Felici, A., Lanfranchi, M., Lanterna, G., Macherelli, A. and Salvadori, B., 'Preconsolidation of pictorial layers in frescoes', *Il Consolida-mentodegli Apparati Architettonici e Decorativi*, Arcadia Ricerche, Bressanone, 2007.

Dei, L., Radicati, B. and Salvadori, B., 'Sperimentazione di unconsolidante a base di idrossido di calcionanofasicosugliaffreschidella Cappella del Podestà al Museo del Bargello di Firenze: aspettichimico-fisici e prove di colore' [Study on a consolidation treatment based on nano-phase calcium hydroxide …], *Sulle Pitture Murali*, Arcadia Ricerche, Bressanone, 2005.

Deloye, F. X., 'Le calcul minéralogique: application aux monuments anciens,' *Bulletin de Liaison des Laboratoires des Ponts et Chaussées*, No. 175, 1991.

DETR, 'The thaumasite form of sulphate attack: Risks, diagnosis, remedial works and guidance on new construction', *Report of the Thaumasite Expert Group*, Department of the Environment, Transport and the Regions, London, 1999.

Dibdin, W. J., *The Composition and Strength of Mortars*, The Royal Institute of British Architects, London, 1911.

Dictionary of Architecture, Vols 2 and 7, The Architectural Publication Society, London, 1853–92.

Dimitriadis, K., 'The ground penetrating radar method used as a tool for monument restoration: Proposed usage and limitations in the detection of stone damage', presented at the Stonecore conference, 'Stone Conservation for refurbishment of Buildings', 23 August 2011, Freiberg, Germany.

Dobson, E., *Rudiments of the Art of Building*, John Weale, London, 1854.

Donaldson, T. L., *Lime, mortar, stucco, and cement: being an article, headed stucco, in the volume of miscellanies in the Encyclopedia metropolitana*, Benjamin Fellowes, London, 1840.

Donaldson, T. L., *Encyclopaedia Metropolitana*, Vol. XXV, London, 1845.

Donaldson, T. L., *Handbook of Specifications, Parts 1 & 2*, Atchley, London, 1859.

Drdacky, M., and Slizkova, Z., 'Calcium hydroxide based consolidation of lime mortars and stone', *Stone Consolidation in Cultural Heritage*, LNEC, Lisbon, 2008.

Dron, R., Brivot, F, 'Bases minéralogiques de sélection des pouzzolanes', *Bulletin de Liaison des Laboratoires des Ponts et Chaussées*, No. 92, 1977.

Durnan, N., 'Wells Cathedral: West Front 2002: Report on the re-treatment of statues', *Journal of the Building Limes Forum*, 2003, Volume 10.

Eckel, E. C., *Cements, Limes and Plasters*, Donhead Publishing Ltd, Shaftesbury, 2005.

Ellis, P. R., 'Analysis of mortars (to include historic mortars) by differential thermal analysis', *Proceedings of RILEM International Workshop PRO12*, Paisley, 2000.

English Heritage, *The English Heritage Directory of Building Limes,* Donhead Publishing Ltd, Shaftesbury, 1997.

English Heritage, *The English Heritage Directory of Building Sands and Aggregates,* Donhead Publishing Ltd, Shaftesbury, 2000.

English Heritage, *Practical Building Conservation: Mortars, Renders & Plasters*, Ashgate Publishing Ltd, Farnham, 2012.

English Heritage, *Practical Building Conservation: Stone*, Ashgate Publishing Ltd, Farnham, 2012.

Everett, A., *Mitchell's Building Construction*, B. T. Batsford Limited, London, 1978.

Everett, A., revised Barritt, C., *Mitchell's Materials*, 5th edition, Longman Scientific and Technical, Harlow, 1994.

Faure, B. and Geoffray, J. M., 'Les basaltes – planches expérimentales basalte-chaux', *Bulletin de Liaison des Laboratoires des Ponts et Chaussées*, No. 94, 1978.

Fenart-Cuvelier, M., *L'age d'or des fours à chaux et cimentiers*, [The golden age of lime kilns and cement], collection of old documents and postcards, Basse Bretagne, Queven, 1991.

Forster, A., 'Hot lime mortars: A current perspective', *Journal of Architectural Conservation*, Vol. 10, No. 3, 2004.

Fourcroy de Ramecourt, M., *L'Art du Chaufournier*, l'Académie Royale des Sciences & Arts de Metz, 1766.

Fournier, A., 'Les facteurs de qualité des chaux industrielles', *Bulletin de Liaison des Laboratoires des Ponts et Chaussées*, No. 79, 1975.

Fournier, M. and Geoffray, J. M., 'Les liant pouzzolane chaux', *Bulletin de Liaison des Laboratoires des Ponts et Chaussées*, No. 93, 1978.

Geoffray, J. M. and Valladeau, R., 'Traitement des sables alluvionnaires par le liant pouzzolanique', *Bulletin de Liaison des Laboratoires des Ponts et Chaussées*, No. 93, 1978.

Gibbons, P., *Preparation and Use of Gauged Hot Lime Mixes*, unpublished paper, 2000.

Gibbons, P., *Preparation and Use of Lime Mortars*, Scottish Lime Centre. Historic Scotland, Edinburgh, 2003.

Gibbons, P. and Leslie, A., Historic Scotland Technical Advice Note 19: *Scottish Aggregates for Building Conservation,* Historic Scotland, Edinburgh, 1999.

Giorgio, R., Dei, L. and Baglioni, P., 'A new method for consolidating wall paintings based on dispersions of lime in alcohol', *Studies in Conservation*, Vol. 45, No. 3, 2000.

Gourdin, W. H. and Kingery, W. D., 'The beginnings of pyrotechnology: Neolithic and Egyptian lime plaster', *Journal of Field Archaeology*, Vol. 2, No. 2, 1975.

Grant, J., *Experiments on the Strength of Cement*, F. & F. N. Spon, London, 1875.

Grizzard, F. J., 'Documentary History of the Construction of the Buildings at the University of Virginia, 1817–1828', PhD thesis, 1996, see Appendix T http:// etext.virginia.edu/jefferson/grizzard/appt.html (accessed 19 September 2006).

Groot, C. J. W., Bartos, P. J. M. and Hughes, J. J., 'Historic mortars: Characteristic and tests – Concluding summary and state-of-the-art', *Proceedings of RILEM International Workshop PRO12*, Paisley, 2000.

Gwilt, J., *An Encyclopaedia of Architecture*, revised by Wyatt Papworth, J. B., Longmans Green & Co., London and New York, 1894.

Gwynn, J., *London and Westminster Improved*, Gregg International Publishers, Farnborough, facsimile of the 1766 edition, 1969.

Henry, A. (ed), *Stone Conservation*, Donhead Publishing Ltd, Shaftesbury, 2006.

Hewlett, P. C. and Lea, F. (eds.), *Lea's Chemistry of Cement and Concrete*, 4th edition, Arnold, London, 1998.

Higgins, B., *Experiments and observations made with the view of improving the art of composing and applying calcareous cements, and of preparing quick-lime: … and specification of the author's cheap and durable cement for building, incrustation, or stuccoing, and artificial stone*, T. Cadell, London, 1780.

Hill, N., Holmes, S. and Mather, D., *Lime and Other Alternative Cements*, Intermediate Technology Publications, London, 1992.

Hirst Conservation, *Condition survey of the fourteenth century wall paintings in All Saints Church, Little Kimble, Buckinghamshire*, revised report dated August 2009.

Historic Scotland, *Memorandum of guidance on listed buildings and conservation areas*, Historic Scotland, Edinburgh, 1998.

Historic Scotland, *Technical Advice Note 19 – Scottish Aggregates for Building Conservation*, Historic Scotland, Edinburgh, 1999.

Historic Scotland, *Mortars in Historic Buildings: A Review of the Conservation, Technical and Scientific Literature*, Historic Scotland, Edinburgh, 2003.

Holmes, S., 'Hot lime in a cold climate', *Journal of the Building Limes Forum*, Vol. 1, No. 2, 1993.

Holmes, S., 'Small scale lime production, hydraulic mortars, classification and standards', *Lime News*, Vol. 5, 1997.

Holmes, S., 'To wake a gentle giant – grey chalk limes test the standards', *Journal of the Building Limes Forum*, Vol. 13, 2006.

Holmes, S. and Wingate, M., *Building with Lime*, Intermediate Technology Publications, London, 1997.

Howe, E., *A report on the conservation of the Lichfield Angel*, report dated May 2007.

Howe, J. A., *The Geology of Building Stones*, Edward Arnold, London, 1910, reprinted by Donhead Publishing Ltd, Shaftesbury, 2001.

Howell, J., 'On the level; plastered', *Daily Telegraph*, 12 June 2002.

Hughes, D. C., et al., 'Calcination of marls to produce Roman Cement', ed. Edison, M. P. *Natural Cement STP 1494*, ASTM, West Conshocken, PA, 2008.

Hughes, D. C., et al., 'Roman cements – belite cements calcined at low temperature', *Cement and Concrete Research*, Vol. 39, 2009.

Hughes, J. J. and Leslie, A. B., 'The petrography of lime inclusions in historic lime based mortars', *Proceedings of the 8th Euroseminar on Microscopy Applied to Building Materials*, Athens, 2001.

Illston, J. M. and Domone, P. L. J., *Construction Materials: their Nature and Behaviour*, 3rd edition, E. & F. N. Spon, London, 2001.

Induni, B. Induni, E., *Using Lime*, Taunton, Somerset, 1990.

Ingham, J., 'Laboratory investigation of lime mortars, plasters and renders', *Journal of the Building Limes Forum*, Vol. 10, 2003.

Ingham, J. P., 'Investigation of traditional lime mortars – the role of optical microscopy', *Proceedings of the 10th Euroseminar on Microscopy Applied to Building Materials*, Paisley, June 2005.

Ingham, J. P., 'The role of light microscopy in the investigation of historic masonry structures', *Proceedings of the Royal Microscopical Society*, Vol. 40/1, 2005.

Ingham, J. P., 'Forensic engineering of fire-damaged structures', *Proceedings of the Institution of Civil Engineers, Civil Engineering*, 162, Special issue – Forensic engineering, May 2009.

Ingham, J. P., *Geomaterials under the microscope – a colour guide*, Manson Publishing Ltd, London, 2011.

Jedrzejewska, H., 'New methods in investigation of ancient mortars', *Archaeological Chemistry*, papers from the Symposium on Archaeological Chemistry, American Chemical Society, Washington D.C., 1967.

Kingery, W. D., Vandiver, P. D., and Prickett, M., 'The beginnings of pyrotechnology, Part II: Production and use of lime and gypsum plaster in the Pre-Pottery Neolithic Near East', *Journal of Field Archaeology*, Vol. 15, No. 2, 1988.

Lambert, P., and Rieu, R., 'Pouzzolane Chaux en technique routière', *Bulletin de Liaison des Laboratoires des Ponts et Chaussées*, No. 93, 1978.

Largent, R., 'Estimation de l'activité pouzzolanique – recherche d'un essai', *Bulletin de Liaison des Laboratoires des Ponts et Chaussées*, No. 93, 1978.

Leslie, A. B. and Hughes, J. J., 'Binder microstructure in lime mortars: implications for the interpretation of analysis results', *Quarterly Journal of Engineering Geology and Hydrology*, Vol. 35, 2002.

Lewin, S. Z., 'X-Ray Diffraction and Scanning Electron Microscope Analysis of Conventional Mortars', *Mortars, Cements and Grouts Used in the Conservation of Historic Buildings*, ICCROM, Rome, proceedings of a conference held 3–6 November 1981, 1982.

Lindqvist, J. E., 'Petrografisk analys av kalksten från Kakeled (Petrografic Analysis of Limestone from Kakeled)', unpublished report, SP Technical Research Institute of Sweden, 2005.

Lithgow, R., *Revised preliminary survey of wall paintings, All Saint's Church, Little Kimble*, report dated 25 January 2001.

Livesey, P., 'Portland cement properties through the ages,' *Journal of the Building Limes Forum*, Vol. 10, 2003.

Lynch G. C. J., *Brickwork: History, Technology and Practice*, Donhead Publishing Ltd, Shaftesbury, 1994.

Lynch G. C. J., 'Lime mortars for brickwork: Traditional practice and modern misconceptions', *Journal of Architectural Conservation*, Vol. 4, Nos 1 and 2, March and July 1998.

Lynch G. C. J., *Gauged Brickwork: A Technical Handbook*, Donhead Publishing Ltd, Shaftesbury, 2006.

Lynch, G., 'Lime mortars – The myth in the mix,' *Building Conservation Directory*, Cathedral Conservation, Tisbury, 2007.

McKay, W. B., *Building Construction*, Longmans, Green and Co., London, 1938–1952, reprinted by Donhead Publishing Ltd, Shaftesbury, 2005.

McKibbins, L. D., Melbourne, C., Sawar, N., and Sicilia Gaillard, C., 'Masonry arch bridges: condition appraisal and remedial treatment', *CIRIA C656*, CIRIA, London, 2006.

Middendorf, B., 'Physio-mechanical and microstructural characteristic of historic and restoration mortars based on gypsum: Current knowledge and perspective', *Natural Stone, Weathering Phenomena, Conservation Strategies and Case Studies*, eds. Siegesmund, S., Weiss, T. and Vollbrecht, A., Geological Society, London, Special Publications 205, 2002.

Millar, W., *Plastering Plain and Decorative*, 1897, reprinted by Donhead Publishing Ltd, Shaftesbury, 1998.

Millet, J. M. and Hommey, R., 'Etude minéralogiques des pâtes pouzzolane', *Bulletin de Liaison des Laboratoires des Ponts et Chaussées*, No. 74, 1974.

Millet, J. M., Hommey, R. and Brivot, F., 'Dosage de la phase vitreuse dans les matériaux pouzzolaniques', *Bulletin de Liaison des Laboratoires des Ponts et Chaussées*, No. 92, 1977.

Millet, J. M., Fournier, A. and Sierra, R., 'Rôle des chaux industrielles dans leurs emplois avec les matériaux à caractère pouzzolanique', *Bulletin de Liaison des Laboratoires des Ponts et Chaussées*, No. 83, 1976.

Mitchell, J., *Geological Researches Around London*, Vol. 1, Manuscript of the Geological Society Library, London, undated.

Mueller, U. and Hansen, E. F., 'Use of digital image analysis in conservation of building materials', *Proceedings of the 8th Euroseminar on Microscopy Applied to Building Materials*, Athens, 2001.

Munsell™ Soil Color Charts, Revised Edition, Macbeth Division of Kallmorgan Instruments Corporation, USA, 1994.

Neve, R., *The City and Country Purchaser and Builders Dictionary*, Sprint, Rivington and others, London, 1736.

Neville, A. M. and Brooks, J. J., *Concrete Technology*, Longman Publications, Harlow, 1993.

Nicholson, P., *An Architectural and Engineering Dictionary*, Vol. 1, John Weale, London, 1835.

Nicholson, P., *The New Practical Builder*, Thomas Kelly, London, 1823.

North, F. J., *Limestones: Their Origins, Distribution and Uses*, Thomas Murby & Co., London and New York, 1930.

Oates, J. A. H., *Lime and Limestone: Chemistry and Technology, Production and Uses*, Wiley-VCH, Germany, 1998.

Palmer, J. J., 'Whose lime is it anyway?', *Natural Stone Specialist*, Vol. 33, 1998.

Papworth, W., *The Dictionary of Architecture*, Architectural Publications Society, London, 1892.

Pasley, C. W., *Observations on Limes*, 1838, reprinted by Donhead Publishing Ltd, Shaftesbury, 1997.

Pasley, C. W., *Observations, deduced from experiment, upon the natural water cements of England, and the Artificial Cements that may be used as substitutes for them*, Establishment for Field Instruction, Chatham, 1830.

Pevsner, N., *The Buildings of England: Buckinghamshire*, Penguin Books, Harmondsworth, reprinted 1979.

Pichon, H., Gaudon, P., Benhassaine, A., Eterradossi, O., 'Caractérisation et quantification de la fraction réactive dans les pouzzolanes volcaniques', *Bulletin – Laboratoires Des Ponts Et Chaussées*, No. 201, 1996.

Prevost, J., *Le Ciment de Vassy: Les Travaux en Ciment*, Société Anonyme des Ciments de Vassy, Paris, 1906.

Quarry Management, *Directory of Quarries & Quarry Equipment (32nd edition)*. QMJ Publishing, Nottingham, 2009/2010.

RILEM, 'Rilem TC 167-Con: Characterisation of old mortars', RILEM Draft Recommendation, *Materials and Structures*, Vol. 34, 2001.

Rivington, N., *Series of Notes on Building Construction* Part III *Materials*, 4th edition, Longmans Green & Co., London, New York and Bombay, 1899, reprinted as *Rivington's Building Construction*, Donhead Publishing Ltd, Shaftesbury, 2003.

Rothstein, E. E. von, *Handledning i allmäna byggnadsläran (Manual for Building Instruction)*, 3rd edition, Stockholm, 1856, 1873, 1890.

Salzman, L. F., *Building in England down to 1540*, Clarendon Press, Oxford, 1952.

Sandström Malinowski, E., Interim reports: PS-3, PS-4, PS-5, PS-7 (Unpublished, 2004–09).

Sandström Malinowski, E., 'Rediscovering historic mortars and skills at Läckö Castle', *Journal of the Building Limes Forum*, Vol. 12, 2005.

Sandström Malinowksi, E., 'Historic mortars revived: Developing local materials and crafts for restoration', *Repairs Mortars for Historic Masonry*, RILEM Publications SARL, 2009.

Sandström Malinowski, E., Historic Mortars at Läckö Castle (in prep, 2011).

Schofield, J., *Lime in Building*, Black Dog Press, Cullompton, 1994.

Searle, A. B., *Limestone and its Products*, Ernest Benn Limited, London, 1935.

Seir Hansen, T., *Research project reports, Microscopy of thin sections – contemporary mortars*, 2005–09.

Shore, B. C. G., *Stones of Britain: A Pictorial Guide to Those in Charge of Valuable Buildings*, Leonard Hill (Books) Ltd, London, 1957.

Siddall, R., 'The use of volcaniclastic material in Roman hydraulic concretes: a brief review', *The Archaeology of Geological Catastrophes*, eds. McGuire, W. G., Griffiths, D. R., Hancock, P. L., and Stewart, I. S., Geological Society, London, Special Publications Vol. 171, 2000.

Skempton, Professor A. W., *Portland Cements, 1843–1887*, excerpt from the Transactions of the Newcomen Society, Vol. XXXV, 1962–63.

Smeaton, A. C., *The Builder's Pocket Manual*, M Taylor, London, 1837.

Smeaton, J., *A Narrative of Building and a Description of the Construction of the Eddystone Lighthouse with Stone*, 2nd edition, G. Nicol, London, 1793.

Smeaton, J., *John Smeaton's Diary of his Journey to the Low Countries 1755*, The Newcomen Society, Leamington Spa, 1938.

Smeaton, J., *Reports of the late John Smeaton FRS, made on various occasions in the course of his employment as a civil engineer*, second edition, 2 vols, M. Taylor, London, 1837. (First published in an abridged, limited edition in 3 volumes: volume 1 in 1797 and volumes 1, 2 & 3 in 1812).

Smith, P., *Notes on Building Construction: Part 3 – Materials*, Rivingtons, London, 1875, reprinted by Donhead Publishing Ltd, Shaftesbury, 2004.

Smith, W., Original Geological Map, displayed at Burlington House, London, 1815–2006. (Winchester, S., *The Map that Changed the World*, Viking, London, 2001).

Sommain, D., 'Technical specification: Use of prompt natural cement in mixes with natural hydraulic limes', *The Louis Vicat Technical Centre – Special Binders Section*, 20 June 2006.

Spalding, F. P., *Notes on the Testing and Use of Hydraulic Cement*, Andrus, Ithaca, New York, 1893.

St John, D. A., Poole, A. B. and Sims, I., *Concrete Petrography, a Handbook of Investigative Techniques*, Edward Arnold, London, 1997.

Summerson, J., *Georgian London*, Pleiades Books, London, 1945.

Swann, S., 'Castle House, Bridgwater, Somerset, Conservators Report for the North and West facades', *Report to SAVE*, October 2003.

Talero, R., 'Kinetochemical and morphological differentiation of ettringites by the Le Chatelier–Anstett test', *Cement and Concrete Research*, Vol. 32, 2002.

Taylor, H. F. W., *Cement Chemistry*, Academic Press, London, 1990.

Telford, T., *A copy of a letter to the Secretary of the British Society*, 1796.

The Institution of Structural Engineers, *Appraisal of existing structures*, (2nd edition), SETO Ltd, London, 1996.

Teutonico, J. M., McCaig, I., Burns, C. and Ashurst, J., 'The Smeaton Project: Factors affecting the properties of lime-based mortars', APT Bulletin, Vol. 25, No.3/4, Association for Preservation Technology International.

Tristram, E. W., *English Wall Painting of the Fourteenth Century*, Routledge & Paul, London, 1955.

Valek, J., Hughes, J. J. and Bartos, P. J. M., 'Portable probe gas permeability in the testing of historic masonry and mortars,' *Proceedings of RILEM International Workshop PRO12*, 2000.

Varas, M. J., et al., 'Natural Cement as the precursor of Portland Cement: Methodology for its Identification', *Cement and Concrete Research*, Vol. 35, 2005.

Verhelst, F., Kjaer, E., Jaeger, W., Middendorf. B., van Balen, K. and Walker, P., 'Masonry – sustainable, contemporary and durable: Anachronism, bold statement or visionary outlook?', *Mauerwerk*, Vol. 15, No. 2, 2011.

Vicat, L. J, *Recherches statistiques sur les substances calcaires à chaux hydraulique et à ciments naturels*, Carilian-Goeury et Vor. Dalmont, Paris, 1853.

Vicat, L. J., *Mortars and Cements*, trans by Smith, 1837, reprinted by Donhead Publishing Ltd, Shaftesbury, 1997.

Vicat, L. J., *Recherches expérimentales sur les chaux de construction, les betons et les mortiers ordinaires*, chez Goujon, libraire de LL. AA. RR. Mme La Duchesse De Berry et Mme La Duchesse D'Orléans, Paris, 1818.

Vitruvius, *De Architectur c.* 15 BC.

Voinovitch, I. A., et al., 'Analyse rapide des ciments', *Chem. Anal.*, Vol. 50, 1968.

Vyskocilova, R., et al., 'Hydration processes in pastes of several natural cements', ed. Edison, M. P., *Natural Cement STP 1494*, ASTM, West Conshocken, PA, 2008.

Wingate, M., *An Introduction to Building Limes*, Information Sheet 9, SPAB, London, 1989.

Yates, T. and Ferguson, A., *NHBC Research Paper NF12: The use of lime-based mortars in new build*, NHBC Foundation, Milton Keynes, 2008.

Zhou, Z., Walker, P. and d'Ayala, D., 'Strength characteristics of hydraulic lime mortared brickwork,' *ICE Proceedings: Construction Materials*, Vol. 161, No. 4, November 2008.

Ziegenbalg, G., 'Colloidal calcium hydroxide: a new material for consolidation and conservation of carbonatic stones', 11th International Congress on Deterioration and Conservation of Stone, Torun, Poland, 2008.

Ziegenbalg, G., Bruemmer, K. and Pianski, J., 'Nano-lime – a new material for the consolidation and conservation of historic mortars', *2nd Historic Mortars Conference*, RILEM, Prague, 2010.

Index

Note: page numbers in italics refer to figures.

For Product Safety Concerns and Information please contact our EU
representative GPSR@taylorandfrancis.com Taylor & Francis Verlag GmbH,
Kaufingerstraße 24, 80331 München, Germany

Printed and bound by CPI Group (UK) Ltd, Croydon, CR0 4YY
01/05/2025
01858481-0001